AF844325

The Developing Role of Public Libraries in Emergency Management:

Emerging Research and Opportunities

Michael Mabe
Chesterfield County Public Library, USA

Emily A. Ashley
Chesterfield County, USA

A volume in the Advances in Library and Information Science (ALIS) Book Series

IGI Global
DISSEMINATOR OF KNOWLEDGE

www.igi-global.com

Published in the United States of America by
IGI Global
Information Science Reference (an imprint of IGI Global)
701 E. Chocolate Avenue
Hershey PA 17033
Tel: 717-533-8845
Fax: 717-533-8661
E-mail: cust@igi-global.com
Web site: http://www.igi-global.com

Copyright © 2017 by IGI Global. All rights reserved. No part of this publication may be reproduced, stored or distributed in any form or by any means, electronic or mechanical, including photocopying, without written permission from the publisher.
Product or company names used in this set are for identification purposes only. Inclusion of the names of the products or companies does not indicate a claim of ownership by IGI Global of the trademark or registered trademark.

Library of Congress Cataloging-in-Publication Data

Names: Mabe, Michael, 1956- author. | Ashley, Emily A., 1981- author.
Title: The developing role of public libraries in emergency management : emerging research and opportunities / by Michael Mabe and Emily A. Ashley.
Description: Hershey, PA : Information Science Reference, [2017] | Includes bibliographical references and index.
Identifiers: LCCN 2016056313| ISBN 9781522521969 (hardcover) | ISBN 9781522521976 (ebook)
Subjects: LCSH: Libraries and community--United States. | Emergency management--United States. | Public libraries--United States--Administration. | Library planning. | Library buildings--Safety measures.
Classification: LCC Z716.4 .M29 2017 | DDC 021.2--dc23 LC record available at https://lccn.loc.gov/2016056313

This book is published in the IGI Global book series Advances in Library and Information Science (ALIS) (ISSN: 2326-4136; eISSN: 2326-4144)

British Cataloguing in Publication Data
A Cataloguing in Publication record for this book is available from the British Library.

All work contributed to this book is new, previously-unpublished material. The views expressed in this book are those of the authors, but not necessarily of the publisher.

Advances in Library and Information Science (ALIS) Book Series

ISSN:2326-4136
EISSN:2326-4144

Mission

The **Advances in Library and Information Science (ALIS) Book Series** is comprised of high quality, research-oriented publications on the continuing developments and trends affecting the public, school, and academic fields, as well as specialized libraries and librarians globally. These discussions on professional and organizational considerations in library and information resource development and management assist in showcasing the latest methodologies and tools in the field.

The **ALIS Book Series** aims to expand the body of library science literature by covering a wide range of topics affecting the profession and field at large. The series also seeks to provide readers with an essential resource for uncovering the latest research in library and information science management, development, and technologies.

Coverage

- Embedded Librarianship
- Ethical Practices in Libraries
- Joint-Use Libraries
- Managing Libraries
- Digital Libraries
- Public Library Funding
- Professional Identity of Librarians
- Outsourcing of Library Services
- Paraprofessionals in Libraries
- University Libraries in Developing Countries

IGI Global is currently accepting manuscripts for publication within this series. To submit a proposal for a volume in this series, please contact our Acquisition Editors at Acquisitions@igi-global.com or visit: http://www.igi-global.com/publish/.

The Advances in Library and Information Science (ALIS) Book Series (ISSN 2326-4136) is published by IGI Global, 701 E. Chocolate Avenue, Hershey, PA 17033-1240, USA, www.igi-global.com. This series is composed of titles available for purchase individually; each title is edited to be contextually exclusive from any other title within the series. For pricing and ordering information please visit http://www.igi-global.com/book-series/advances-library-information-science/73002. Postmaster: Send all address changes to above address. Copyright © 2017 IGI Global. All rights, including translation in other languages reserved by the publisher. No part of this series may be reproduced or used in any form or by any means – graphics, electronic, or mechanical, including photocopying, recording, taping, or information and retrieval systems – without written permission from the publisher, except for non commercial, educational use, including classroom teaching purposes. The views expressed in this series are those of the authors, but not necessarily of IGI Global.

Titles in this Series

For a list of additional titles in this series, please visit:
http://www.igi-global.com/book-series/advances-library-information-science/73002

Library and Information Services for Bioinformatics Education and Research
Shri Ram (Thapar University, India)
Information Science Reference • ©2017 • 291pp • H/C (ISBN: 9781522518716) • US $190.00

Managing Knowledge and Scholarly Assets in Academic Libraries
Bhojaraju Gunjal (National Institute of Technology Rourkela, India)
Information Science Reference • ©2017 • 381pp • H/C (ISBN: 9781522517412) • US $180.00

Interdisciplinary Digital Preservation Tools and Technologies
Tariq Ashraf (University of Delhi, India) and Naresh Kumar (American Center Library, India)
Information Science Reference • ©2017 • 281pp • H/C (ISBN: 9781522516538) • US $190.00

Library Technology Funding, Planning, and Deployment
Edward Iglesias (Stephen F. Austin State University, USA)
Information Science Reference • ©2017 • 257pp • H/C (ISBN: 9781522517351) • US $155.00

Academic Library Development and Administration in China
Lian Ruan (Illinois Fire Service Institute at the University of Illinois at Urbana-Champaign, USA) Qiang Zhu (Peking University, China) and Ying Ye (Nanjing University, China)
Information Science Reference • ©2017 • 391pp • H/C (ISBN: 9781522505501) • US $195.00

Handbook of Research on Emerging Technologies for Digital Preservation...
Alfonso Ippolito (Sapienza University of Rome, Italy) and Michela Cigola (University of Cassino and South Latium, Italy)
Information Science Reference • ©2017 • 649pp • H/C (ISBN: 9781522506805) • US $275.00

Information Seeking Behavior and Challenges in Digital Libraries
Adeyinka Tella (University of Ilorin, Nigeria)
Information Science Reference • ©2016 • 359pp • H/C (ISBN: 9781522502968) • US $185.00

For an enitre list of titles in this series, please visit:
http://www.igi-global.com/book-series/advances-library-information-science/73002

IGI GLOBAL
DISSEMINATOR OF KNOWLEDGE

www.igi-global.com

701 East Chocolate Avenue, Hershey, PA 17033, USA
Tel: 717-533-8845 x100 • Fax: 717-533-8661
E-Mail: cust@igi-global.com • www.igi-global.com

Table of Contents

Preface

In the fall of 2011, Hurricane Irene, the ninth-named storm of the Atlantic hurricane season, turned out to be more powerful than many communities expected (Hazard, 2003). Hurricane Irene arrived in Chesterfield County beginning early Saturday morning August 27, 2011, and continued through early Sunday morning August 28, 2011. Because of their newly assigned responsibilities with volunteer coordination and donation management, Chesterfield County Public Library (CCPL) leadership attended the Emergency Operations Center (EOC) daily briefings. While Irene was not initially expected to make a significant impact, county leaders were active, cautious, and prepared. In the immediate aftermath of the storm, county leaders reported 799 roads were reported blocked by storm damage, over 70% of the county was without power, 232 homes were damaged and there was one fatality. Locally there was $1.7 million in damage to public facilities and $3.8 million in private property damage. In the end 56 deaths and approximately $15.6 million in damages were incurred throughout the path of the storm. (Cloak, 2012)

During these briefings emergency management reports delivered by key players reported early on that Irene was, in fact, worse than expected. During the open discussion, I informed the group that several thousand customers showed up the day before at eight of the local branch libraries seeking damage reports and storm and power outage updates as well as access to restroom facilities for personal care and internet access to communicate with employers, family and friends. My initial report was new to the EOC discussion. Many dismissed the report as an outlier and I was not surprised at their reaction.

During the second day's briefing, I was called upon by the Emergency Management Coordinator, Emily Ashley, to report on activities at the library since the last briefing. Without hesitation, I reported the customer door count, which was similar to the previous day, and relayed several customer comments that indicated the library had become their lifeboat during this event. Ms.

Ashley eventually made it clear to the EOC team that she believed the fact that so many county residents visiting the library for services was a critical piece of information in developing solutions for mass care needs and she would continue to include a daily report from the library.

With a large portion of Chesterfield County without power, concerns with activating the Emergency Support Function (ESF) mass care began to be discussed during EOC briefings. As described in the office report submitted by CCPL administration (Mabe, 2016) during one particular EOC briefing, the emergency management director broached the subject of how to handle mass feeding with human services division staff and leaders who were responsible for this particular task. I volunteered any branch of the library and any of its staff for mass care feeding duty. Social Services staff was a little hesitant to accept the offer of help. They expressed concern that the volume of residents who may come to the library for mass care service might be too much for the staff and the facility to handle. A discussion ensued about the volume of customers who regularly use the library on a daily basis. A comparison was also drawn between checking out books vs. handing out prepackaged meals. Once the connections were made and the picture became clearer, libraries were enlisted in the mass care process as feeding stations.

Within 24 hours, Red Cross vehicles were dispatched to the LaPrade Branch Library and word was distributed to the needy communities that meals were available at the library. Over the next four days 2,000 meals were distributed at the LaPrade Branch library at predetermined times from the Red Cross truck. Freeze dried, Meals Ready to Eat (MRE) were also left at the LaPrade Branch and a few CCPL locations to be distributed by staff across the circulation desk for those residents who could not meet the Red Cross truck schedule. With so many visitors staying through the daytime as reported during EOC briefings, emergency management leaders assigned law enforcement officers to coordinate traffic and crowd control at the branches.

Although CCPL had been offering services in support of disaster victims before and after emergencies for years, it was not until Chesterfield County's Emergency Management Coordinator Emily Ashley accepted my offer as library director to provide a location and staff resources in support of mass care feeding that CCPL's emergency management collaboration was recognized. Ms. Ashley's willingness to listen to my recommendation was a first. She clearly saw the value of the library facility and the library staff as a key to solving the immediate mass care feeding problem she was responsible for solving. After successfully completing this task, Ms. Ashley and I found a common ground in support of her department's mission of mounting an effective response and the library's emerging role during emergencies. Since 2011 Ms. Ashley and I continue to coordinate efforts in several real emergencies and mock training exercises. CCPL's current success in support of emergency management is based on the strong relationship that now exists between the public library and the county department of emergency management.

Michael R. Mabe
Chesterfield County Public Library, USA

Acknowledgment

Special thanks to Lori Purcell for her exceptional research skills and her contributions in making the message of this volume clear and precise, Blanche DePonte for her editing and proofreading skills, and Paula Dias for her tireless technical support.

Introduction

The natural role the public library plays in local emergencies has emerged and developed over the years and is unique to each library and community. In more recent times, public libraries have operated as emergency community centers providing life sustaining essentials to victims such as shelter, food and access to communication tools and resources during and after disaster situations. Their size, location, and public amenities such as restrooms, internet access and electrical power, so needed in our online world, makes them a natural for victim relief. Libraries are also locally available and frequently include spacious meeting rooms that can be used as command centers for official and unofficial briefings which make them a natural for disaster support. All combined, these assets make the public library a logistical hub that provides critical access and the delivery of disaster information, housing and food, and finally, skilled librarians who possess critical communication skills, expert knowledge of information and information systems, and a broad understanding of community resources and leaders.

This volume will explore what is emergency management, including describing why public policy and traditional reactionary approaches to emergency management have caused the profession to grow and develop in a different direction. The volume will also describe the fundamental practices used within emergency management and that constitute their operational approach: i.e. prevention, protection, response, recovery, and mitigation. Much of what the emergency management professional does frequently aligns with the day to day activities and skills of the library and the librarian.

To be effective in emergency management, local library leaders need to understand the local emergency management command structure process and comprehend where and how the local public library can fit in that structure. This awareness starts with leaders gaining knowledge of local municipal government structures and developing partnerships with other municipal agencies that provide the support for community mass care needs. It continues with librarians acquiring essential emergency training including specialized Community Emergency Response Training (CERT). This specific training can be seen as an extension of the daily role most libraries fill when connecting people and resources but applied only during emergencies and disasters. The CERT role supports survivors and emergency management professionals with knowledge of and access to first aid, communications, Wi-Fi, local news and municipal data, a place to warm up or cool down and charge a cell phone among other things. The only difference between their daily role and a CERT role is when the service is offered. When librarians are trained properly to respond to a disaster, they can then offer public outreach about the need to prepare personally during periods of non-disaster times. Any action public libraries take ahead of time to prepare for disasters locally will build trust in their community and assure residents that the library is an effective resource to go to when disasters happen.

Current approaches and past experiences of libraries engaging in emergency services fall into five specific areas:

1. Preservation of materials and the safety of staff subjected to disaster situations.
2. Providing normal library services to victims, survivors.
3. Offering specific services due to designated roles in support of disaster needs such as managing volunteers and donations.
4. Providing access to critical information such as storm alerts, power updates, and medical information.
5. Providing unintended services that may help in the recovery of a disaster such as internet services so that victims can file disaster related forms with FEMA.

Before Hurricane Katrina in 2005, libraries limited their role during disasters to preservation and providing routine services after the disaster had ceased. After Katrina and in each subsequent disaster since then, public libraries have begun to play designated roles and have learned to respond in more specific ways in support of first responders and survivors. Since no two disasters are alike, the key to public libraries developing a role in their local community individually and collectively is rooted in the willingness of library leaders and staff to extend their skills, talents, and resources to first responders and survivors, and operating as emergency centers and information hubs beyond normal operating hours before, during and after an emergency event.

Examples of libraries extending that vision have become more pronounced since Hurricane Katrina. Katrina and subsequent disasters provide local communities a laboratory for libraries and emergency professionals to evaluate and analyze their responses for the purpose of improving future responses to emergencies and disasters. The authors of this volume will provide multiple examples of libraries stepping outside their comfort zone and experiencing substantial success, often in first responder style circumstances. The authors' professional experience during Hurricane Irene is provided in more specific detail through the volume as one of many emerging models of the developing role of public library in emergency situations. The Chesterfield County Public Library (CCPL) model, while not as dramatic as some, is a clear example of starting with serendipity and continuing with deliberate action by working with local emergency management officials before, during and after an emergency situation.

All libraries can engage in an emergency response, but not all libraries will be able to offer response at the same level. In preparing to address the developing role of the public library in emergency situations, three key elements need to be considered by local library staff and community leaders in order to determine why and how a public library could or should engage in an emergency response.

- The first factor is the nature of the library facility. Most libraries are built from the ground up and designed to house stacks and accommodate large crowds. This means that the library facility is generally sturdier than most community facilities. During Hurricane Sandy libraries in New York and West Virginia received very little damage allowing library officials to resume business as soon as the storm had passed. The small number of libraries negatively impacted by disasters including the devastation of Hurricane Sandy speaks to the durability of library structures. Local experts should assess and evaluate the structural nature of their local public library facilities to ensure their capacity to endure damage due to disasters typical to their region.
- The second is the disposition and nature of the librarian. Professionally trained or naturally gifted, librarians are not public safety staff or first responders. The librarian's primary role is to provide access to information and deliver information services during the best of times. Recent experience has shown that librarians are often required to respond in a capacity that is critical in the support of first responders and survivors. It has become clear during recent disasters that librarians are naturally prepared to plan for and organize information, events, materials, volunteers, donated goods of all varieties and amply qualified to interact with individuals and crowds in simple transactions as well as complicated discussions.

- Third, the fact that the library is the community service and institution viewed by most local residents as *the* community center. During disasters of all kinds around the country, residents have been known to gravitate to the library when a large scale emergency arises. After a disaster has affected a community the first thing the residents want to do is "get back to normal." While they may not be familiar with the services the library offers during normal times, in the wake of disaster the library becomes the center of normal. Customers tend to see the library as a clearing house of many things regardless of previous customer usage. At the library, victims can receive service from with staff that are helpful and interact with known members of their community. This immediate gratification offers a sense of comfort that can help them in their return to normal.

These fluid ideas of the library, librarianship, and the library as a community center during disasters have been demonstrated by many public libraries in the past 10 years. This work will answer the question why is the library and librarian important in responding to local emergency situations. Our aim is to provide examples of libraries that provided assistance and describe the ways that libraries can become involved in all facets of an emergency and the role libraries can assume during their local emergency situations. We predict that the reader will gain awareness of the background information, previous examples, and practical application between emergency management and the public library. It is the authors' hope that this volume will provide motivating advice and counsel to other public libraries that will help them prepare for when it is their time of need.

Chapter 1
Emergency Management: An Introduction and Brief History

Today's public libraries continue to transcend the historical stereotypes of the past by embracing new roles and responsibilities stemming from the many challenges and opportunities they face in their local community. While their core purpose of providing access to information remains the same, many public libraries and librarians are engaging the community beyond their assumed role and participating in non-traditional activities such as hosting job fairs, aiding small business development, and providing maker spaces. In most of their community engagement, the librarians' current training and skill set supports their expanding role, but not always. For example, supporting local disaster survivors and emergency management professionals before and after a major crisis and during an emergency response, is one area of community engagement which requires a more specialized approach and training. Public librarians are being enlisted to respond more and more often to local disasters.

Specialized knowledge and training will help public libraries become more effective offering the full range of public library resources including staff, equipment, and facilities in support of responses to local emergencies and disasters. The specialized knowledge and training in this process should include:

DOI: 10.4018/978-1-5225-2196-9.ch001

Copyright ©2017, IGI Global. Copying or distributing in print or electronic forms without written permission of IGI Global is prohibited.

- Incorporating effective strategies for pre and post disaster planning and analysis.
- Learning to predict the impact of a disaster and potential response needs.
- Engaging in required partnerships with community and emergency leaders.
- Abiding by the principles of emergency management preparation and training.
- Employing the essential tools and practices to ensure effective participation.

Engaging or participating in emergency response can be as simple as providing computer access or as complex as operating a temporary overnight shelter. To be completely effective, public libraries need direct knowledge and understanding of the principles and practices used by emergency management professionals when mounting an emergency response. This understanding must include a complete knowledge of disaster management history and process, the emergency management command structure, as well as an understanding of the essential disaster response basics and how they are applied.

WHAT IS EMERGENCY MANAGEMENT

Emergency Management has evolved over time. Only recently has emergency management emerged as a professional discipline defined by principles and practices instead of political doctrine and philosophy. For decades disaster management happened as the result of *acts of legislation to render assistance* to individual disasters in the form of financial aid or access to federal government resources. Then in 1950 Congress enacted the Civil Defense Act. This act placed the burden of the civil defense at the state and local level. Although still driven by national policy, the Federal Civil Defense Administration (FDCA) reinforced the belief that disasters were only caused by military incursions and disaster preparation required a military approach. Since that time, emergency management has grown and developed beyond civil defense perceptions.

Today, emergency management professionals realize that a natural disaster is any event that severely disrupts essential community services such as power, water, transportation, and communications. In modern society, communities are not an island unto themselves. The interconnectivity, if interrupted at a critical point of infrastructure within and between the communities separated by any amount of geography, can lead to circumstances that may have disastrous impact.

The disruption of these services at the local level is where a disaster has its greatest effect on the health and economy of local communities, and their residents before and throughout all phases of a disaster. While the mentality has been to think big and respond big, communities have learned through direct experience to prepare and implement natural disaster management plans that manage the effects of a disaster in the local arena.

A HISTORY OF EMERGENCY MANAGEMENT

The emergency management approach to natural disasters of any kind in the United States has been a patchwork of reactionary approaches and strategies that, according to Flynn (2007) always seemed to be looking behind and never ahead. Recent reaction to natural disasters has improved although emergency management practices still seem to be lacking in their capacity to handle the elusive and uncontrollable surprises that are always a part of natural disasters.

The first recorded response to a disaster was in 1803 when Congress provided funds for a town in New Hampshire to recover from an all-consuming destructive fire. Had the occurrence of natural and man-made disasters remained at an 1803 level this type of response would be woefully inadequate today. For more than 130 years there was no obvious change in approach to managing emergencies and disasters including no additional official federal government actions until 1936 when President Franklin Roosevelt implemented the Flood Control Act (Mabe, 2016).

The Flood Control Act of 1936 authorized civil engineering projects such as dams, levees, dikes, and other flood control measures through the United States Army Corps of Engineers and other Federal agencies. The act also dictated that federal investigation and improvements of rivers and other waterways for flood control and allied purposes fall under the jurisdiction of the United States Department of War. According to Arnold (1988), the act concentrated on the subjects of economic development and military preparation rather than emergency management. Unfortunately, the act also unwittingly established the first emergency response doctrine that would dictate emergency management for generations to come.

The rise of the cold war forced congress to seriously consider the needs of the civilian population in the case of nuclear attack. These included the Civil Defense Act and the Federal Disaster Relief Act. These acts placed the burden of civil defense at the state level while at the same time providing a framework to carry out the work required to alleviate the suffering from "major disasters" including flooding. The law required no specific congressional legislation to allocate emergency resources. These acts made access to federal assistance more easily and readily available to the states and encouraged a shift in centralized doctrine from the military preparation to a more effective decentralized doctrine of self-help and local control in managing disasters.

According to Sylves (2015), these acts continued to evolve and grow based on the foreign policy and the military actions of executive leaders with natural disaster response as an afterthought. For example, President Dwight D. Eisenhower used the acts to encourage mass evacuation policy due to military actions. President John F. Kennedy employed the acts to emphasize community volunteerism which President George W. Bush echoed in the aftermath of 9/11. President Lyndon B. Johnson, so preoccupied with the Vietnam War used the acts to deemphasize civil defense all together. It wasn't until a series of nonmilitary natural disasters in 1964/65, Hurricane Betty, the Alaskan Earthquake, and the Indiana tornado caused law makers to stray away from military thought and consider an all hazards or dual use approach to civil defense. This approach directly provided more resources to victims regardless of the cause of the disaster.

President Richard M. Nixon was the first executive to shift from an all military approach. In his Natural Security Decision Memorandum 184, he officially recommended the dual use approach by replacing the Office of Civil Defense with the Defense Civil Preparedness Agency within the Department of Defense. Although still in the Department of Defense, this executive action encouraged Congress to pass the Disaster Relief Act of 1966 allowing the inclusion of threats from natural disasters within the civil defense warning system.

The evolution of the 1950 Civil Defense Act was finally completed in 1978 when Congress amended the original 1950 act to include the dual use approach. The amended act allowed federal dollars previously to be used exclusively for preparing for military attack to also now be used for preparing for natural disasters.

It should be noted that during World War II and the Cold War the federal government implemented the Federal Civil Defense Administration and the Office of Defense Mobilization to react to emergencies. These actions focused more on local civil defense concerns than on natural disasters. Even with additional attempts through the 1960s and 1970s to shift the focus on natural disasters from civil defense to emerging emergency response techniques documented by the Emergency Management Institute (EMI), emergency management continued to be fragmented and ineffective.

In 1979 the Federal Emergency Management Agency (FEMA) was created as a way to be more effectively inclusive in emergency response, regardless of the causes. FEMA's primary purpose is to coordinate the response to a disaster that has occurred in the United States that overwhelms the resources of local and state authorities. The acknowledgement that emergencies must be managed at the local level was an important first step in the development of modern principles and practices for emergency management.

FEMA (2016) learned first-hand the complex yet diverse nature of emergency management with responses to the Love Canal, the Cuban refugee crisis, Three Mile Island nuclear power plant, the Loma Prieta Earthquake in 1989 and Hurricane Andrew in 1992. A former FEMA Director, James L. Witt, who himself was a state emergency manager from Arkansas before he was a FEMA Director, initiated sweeping reforms to the agency beginning in 1993 that streamlined disaster relief and recovery operations. These reforms included a new emphasis on preparedness and mitigation and redirected more of FEMA's limited resources from civil defense into disaster relief, recovery, and mitigation programs.

THE CHANGING NATURE OF DISASTER RESPONSE

In the last several years we have experienced several major disasters that provided a number of communities and local response agencies, including libraries, a laboratory to learn from. Holdeman (2010) notes the decade of disasters from 2000 to 2009 was significant. Beginning in 2000 a great deal of preparation went into preparing computer systems and processes for the transition to a new millennium. There was a widespread concern and even deep fear that computer systems of the world would not be able to update to the millennium and crash. This concern forced many emergency operations centers to open throughout the midnight hours and nervously wait for the system to "shutdown" when 12:00 struck. It is an uncanny reality that because people were in place all throughout the United States and around the world, we were better prepared to deal with that problem than at any other time in history.

In 2001, the terrorist attacks on 9/11 pushed emergency management into a completely new emphasis on terrorism. The formation of the Office of Homeland Security foreshadowed the creation of the new U.S. Department of Homeland Security (DHS). September 11th provided a major communication crisis between first responders and the community in general. With system wide infrastructure shutdown, people did not know how to connect with one another nor did they know where to get help. First responders did not have standardized communication through their ten code language or their radio systems. It was physically and literally impossible for a New York City police officer to talk to a New York City fire fighter. In reaction, George W. Bush's Homeland Security Directives #5 institutionalized the National Incident Management System was instituted as a baseline for first responders. Fifteen years later the practice of NIMS is an intrinsic part of any public safety response which extends to mass care.

In 2004, four hurricanes roared through Florida causing loss of life and billions of dollars of damage. This active Atlantic hurricane season gave Florida emergency management agencies a run for their money. These hurricanes, which crisscrossed the state, nearly bankrupted the National Flood Insurance Program. They also forced us to shift evacuation procedures to only evacuate due to flooding and not due to wind. Post 2004 businesses started to engage in continuity of operations planning with the understanding that fifty percent of small businesses do not reopen post flood or other disaster. A weary Florida developed resiliency plans and rich public outreach campaigns that encouraged hurricane prone areas to develop partnership and push their own communities to do robust public outreach programs.

In 2005, Hurricane Katrina and its impact on New Orleans made the United States look like a third world country. People watched nightly news in shock at the sight of thousands of people standing on their roof tops begging for relief and others that pointed guns at first responders out of fear of the unknown. FEMA Administrator Craig Fugate noted that the biggest problem was that the lengthy decision making process postponed the initial response; therefore, making the horrible situation even worse (Block, 2015). "The trap we fell into was we prepared for what we were capable of doing and just hoped it would never be any worse," added Fugate. "The theory was if it's worse, then we'll just scale up, and we found out scaling up didn't work. So you have to prepare for how bad it's going to be, or you're going to fail. And I think that was the lesson we had to learn in all levels of government" (2015).

This event is considered a decisive moment for those it affected. Meaning, people who suffered through Hurricane Katrina were faced with the very real possibility of dying after having already lost all their possessions. This reality makes people look at disaster response in a very different way. Following Katrina, there was a new emphasis on evacuation plans and the challenges of providing assistance to vulnerable populations before, during, and after a disaster. Similarly, Katrina prompted a renewed emphasis on disaster preparation at the local level.

While Katrina exposed vulnerable populations, little action was taken to correct the problem. Years later the City of Los Angeles was sued to force the use of special needs planning because of their "do the best you can" effort (The Minnesota Governors Council on Developmental Disabilities, n.d.). Los Angeles is notably disaster-prone and has experienced several disasters, from the massive Northridge earthquake in 1994 to the wildfires in 2008 and 2009. Despite this fact, Los Angeles lacks adequate disaster planning for people with disabilities, such as accessible emergency shelters, plans for providing services and medication at shelters, accessible transportation, evacuation assistance, and communication services that are available and accessible to people with a wide range of disabilities. The 2011 court ruling highlighted the problem of people with disabilities, including many seriously ill seniors, were left behind to die because of a lack of disability planning.

In 2006, the death toll of the Sago Mine disaster in West Virginia caught the attention of many people. The unfortunate and ultimately inaccurate report that 13 miners were found alive when 12 of them had actually perished was a reminder to emergency officials everywhere to verify information before making announcements to the media. This incident is also an example of media crossing the line by trying to be the first, rather than being correct. People affected by the disaster stayed holed up in a church waiting on bits and pieces of information. No appointed official assisted with critical incident stress needs or helped the community process information.

In 2007 the Southeastern United States was seized in a multiple-year drought. It was anticipated that many areas of the Southeast would eventually run out of water. Although they were prepared for electrical power outages, the impact of no water for a large metropolitan area of the nation was a daunting hazard to contemplate. Fortunately, that line was never crossed but with climate projections being what they are, this is a hazard that we must prepare for in the future.

According to Sylves (2015), during this same period these experiences caused some changes in emergency management that are of note. In 2002, emergency managers saw the creation of the DHS and the subordination of FEMA into its organizational ranks as a disaster in itself. While DHS controlled the billions, they allocated more in terrorism prevention and response dollars administered by the Department of Justice and less in FEMA for emergency management funding. In 2003, the creation of the mandate to implement the National Incident Management System (NIMS) and a new emphasis on the Incident Command System (ICS). To obtain and retain federal funding, jurisdictions would have to adopt and annually report their progress at becoming NIMS compliant.

While the Fire Service had been missing the Incident Command System since wildfire response in the 1970s, other public safety agencies were grappling with the details of command structure, staging area placement, and phasing out ten code language. The command implementation was vague at best as local governments wrestled with who needed training to be compliant, what constituted an incident commander and did an accountant need to know the same information as an operational field fire chief.

Reporting the details was even more catastrophic as the National Incident Management System Capability and Assessment Tool (NIMSCAST) loosely defined what needed to be reported back to the federal level. Spreadsheets were accounted for with a general fear of being audited at the local level. Funding that was provided had lax guidelines and allowed for spending with no maintenance or sustainment considerations for equipment that would be rarely used and have an expiration date. In many regards the implementation of the program while necessary was in some ways relatable to the wild west of tracking.

In our current decade we have already experienced ominous disasters. The national weather service reports (Wikipedia, The Free Encyclopedia, 2016) that the 2011 Super Outbreak was the largest, costliest, and one of the deadliest tornado outbreaks ever recorded, affecting the Southern, Midwestern, and Northeastern United States and leaving catastrophic destruction in its wake. The event affected Alabama and Mississippi most severely, but it also produced destructive tornadoes in Arkansas, Georgia, Tennessee, and Virginia, and affected many other areas throughout the Southern and Eastern United States.

In total, 362 tornadoes in 21 states from Texas to New York to southern Canada were confirmed by the NOAA's National Weather Service (NWS) and Environment Canada. Widespread and destructive tornadoes occurred each day of the outbreak, with April 27 being the most active day with a record of 218 tornadoes touching down that day from midnight to midnight CDT (0500 – 0500 UTC). Four of the tornadoes were destructive enough to be rated EF5, which is the highest ranking possible on the Enhanced Fujita scale. Typically, EFS tornadoes are only recorded about once each year or less.

In 2012 Hurricane Sandy (unofficially known as "Superstorm Sandy") was the deadliest and most destructive hurricane of the 2012 Atlantic hurricane season, and the second-costliest hurricane in United States history. Classified as the eighteenth named storm, tenth hurricane and second major hurricane of the year, Sandy was a Category 3 storm at its peak intensity when it made landfall in Cuba. While it was a Category 2 storm off the coast of the

Northeastern United States, Sandy became the largest Atlantic hurricane on record (as measured by diameter, with winds spanning 1,100 miles (1,800 km)). Estimates as of 2015 assessed damage to have been about $75 billion (2012 USD), a total surpassed only by Hurricane Katrina. At least 233 people were killed along the path of the storm in eight countries.

EMERGENCY MANAGEMENT TODAY

While Emergency Management grew out of civil defense, due to the Cold War, the changing nature of disasters including the events of 9/11, and Hurricane Katrina redefined the emergency management profession starting with standardized models for action. The NIMS and the ICS model which had lived in the fire service since the 1970s became the baseline standard for Emergency Management. This concept was new to people outside of the para-military fire service. In actuality, the accompanying planning process was systematic and similar to conducting business on a daily basis for emergency professionals. This new dawn of emergency management also led to meeting human services needs beyond the initial property and lifesaving response of short term recovery.

Reaction to natural disasters is generally initiated at the local level with public safety agencies taking the lead. However, more and more local governments are establishing offices of emergency management lead by professional managers who coordinate all emergency responses. For events that exceed local resources, state offices of emergency management generally step in to take charge of coordinating access to resources, although the planning and use of all resources still tends to be managed at the local government level.

FEMA, an agency of DHS, takes the lead on disasters that impact large geographic regions encompassing multiple states or massive destruction. FEMA's emergency management response is administered within ten geographic regions, but does not override state authority or local decision makers.

If a disaster or emergency is declared to be terror related or an "Incident of National Significance," the Secretary of Homeland Security will initiate the National Response Framework (NRF). The NRF allows the integration of federal resources with local, county, state, or tribal entities, with management of those resources still to be handled at the lowest possible level, through the NIMS which provides legal authority. In addition to the national level processes, FEMA recommends mission areas and core capabilities at the local level to ensure that community resources are always available to mount a response at the local level.

MISSION AREAS AND CAPABILITIES OF EMERGENCY MANAGEMENT

A number of emergency management practices involve processing information at the local level. Most emergency managers rely on other departments resources to accept assignments and complete tasks. They do not own any resources that they can instantly deploy or command staff large enough that can analyze information and distribute personnel to assigned staging areas. They maintain an extensive network of partnerships that allow them to run a comprehensive program of collecting and disseminating information to those that do have actual staff and resources assigned to participate in a response.

Depending on the locality, the emergency management program is either a job duty within uniformed public safety, a division within public safety, or a stand-alone department. The goals and objectives remain the same in analyzing data and assigning resources within a standardized designated mission response structure.

Most local emergency management departments strive to cover all aspects of emergency management in their day-to-day and disaster activities in five specific mission areas. The mission areas include Prevention, Protection, Mitigation, Response, and Recovery. These areas support the National Preparedness Goal to make us a secure and resilient nation with the capabilities required across the whole community to prevent, protect against, mitigate, respond to, and recover from the threats and hazards that pose the greatest risk.

The risks include events such as natural disasters, disease pandemics, chemical spills and other man-made hazards, terrorist attacks, and cyber-attacks. To best assist in achieving success in the mission areas, there are thirty-two core capabilities identified in the National Preparedness Goal for all who have a role in emergency management. Not all of the core capabilities are adopted at the local level.

According to Chesterfield County, VA (2016), the core capabilities of planning, public information and warning, and operational coordination, the key to most emergency operations, are already embedded within the five mission areas. The remaining twenty-seven capabilities are particular to one mission area and are best aligned by the local government.

The mission areas and embedded core capabilities used by Chesterfield County, VA are provided as an example only and are described below:

- Prevention comprises the capabilities necessary to avoid, prevent or stop a threatened or actual act of terrorism or man-made disaster. This also includes preparedness measures of conducting a systematic process engaging the whole community as appropriate in the development of executable strategic, operational, or tactical level approaches.

- Protection ensures that the County has the necessary capabilities to secure our residents and infrastructure against acts of terrorism, man-made or natural disasters.
- Mitigation is the effort to reduce loss of life and property by lessening the impact of disasters. This is achieved through risk analysis, which results in information that can be applied to activities that reduce risk pre or post disaster. Mitigation includes the core capability of community resilience that enables the recognition, understanding, communication of, and planning for risk. This mission area is an inclusive whole community approach to empower individuals to make informed decisions to adapt, withstand and quickly recover from incidents that can also reduce long-term vulnerabilities.
- Response comprises the capabilities necessary to save lives, protect property and the environment, and meet the basic human needs after an incident has occurred. This action oriented mission area begins as the disaster occurs. Because all disasters begin locally, in the event an incident exceeds local emergency response capabilities, outside assistance is available, either through mutual support agreements with nearby jurisdictions and volunteer emergency organizations or, through the Virginia Emergency Operations Center (VEOC). A local emergency must be declared and local resources must have intent to be fully committed before state and federal assistance is requested. Response related core capabilities ensure appropriate measures are taken to protect the health and safety of all responders as well as the environment.
- Recovery allows the community to recovery effectively and this may include state and federal partners. Recovery encompasses economic and business activities returning to a healthy state that is defined as an economically viable community for business and employment. When making recovery considerations, the Comprehensive Plan will be used as a tool through Housing and also the protection and natural of cultural resources.

To reiterate, it is through the mission areas that FEMA (2015) suggests thirty-two more specific, core capabilities are administered in achieving the goals of national preparedness. Not all of the capabilities are adopted at the local level. For purposes of this text we will not review the core capabilities and for clarity will only discuss the five mission areas which include the capabilities of planning, public information and warning, and operational coordination, as listed in our definitions above.

Emergency Support Functions are the alignment of the specific roles and duties within an organization's Emergency Operations Plan. Through FEMA there is a standardized grouping of functions to provide support at the federal, state, or local level. While FEMA sets the Emergency Support Function (ESF) structure in the National Response Framework both the State and Local Emergency Operations Plan follow the same structure. By having the 15 specific ESFs, incident command is allowed to be followed, meaning no matter who is responding or shows up to offer assistance, the plans will have a familiarity and continuity to them. The ESFs are mechanisms that allow frequent grouping of people and resources. They are:

1. Transportation
2. Communications
3. Public Works and Engineering
4. Firefighting
5. Emergency Management/ Direction and Coordination
6. Mass Care/Human Services
7. Logistics
8. Public Health and Medical Services
9. Search and Rescue
10. Oil and Hazardous Materials Response
11. Agriculture and Natural Resources
12. Energy
13. Public Safety and Security
14. Long Term Community Recovery
15. External Affairs

Each of these titles has a coordinating agency or department that is responsible for the management oversight upon activation. While the ESF is not considered to be a detailed operational procedure it does give an overview of what human capital and resources will be required. By having this structure the ESF lead is able to manage mission assignments and collaborate with emergency management to provide operational priorities and situational awareness.

In each of the ESFs each coordinating agency is responsible for:

- Participation in the Emergency Operations Center to assist with the development of incident management objectives throughout the planning period
- Providing staff, equipment or other resource support as requested by the Primary Agency or incident commander of the disaster
- Assist in providing information to complete a situational assessment
- Maintain trained personnel to be support emergency response

A quick glance of the listing of each of the ESFs, libraries are able to offer assistance to several areas; however, the most natural fit is into ESF #6- Mass Care. The actual definition of Mass Care is to address the non-medical mass care, emergency assistance, housing and human services needs of indivduals and/or families impacted by natural or human-caused disasters (www.vaemergency.gov/wp-content/uploads/drupal/COVEOP_2012_ESF_6_Mass_Care)

Through this general tasking library staff assist in roles that they perform in their normal job. As the text has shown, librarians provide human service assistance on a daily basis. While not providing temporary shelter assistance libraries are capable of providing a temporary refuge for disaster victims to gain information, connect thorough technology, and plug in devices to name a few critical examples. By purposefully adding training that is directed to emergency management they are enhancing their readiness skills to accept disaster victims. If a library is not already included in the planning process of the EOP or with local emergency management they can begin the partnership by offering assistance through Mass Care. Human service needs are not only met by the staff of the library; but, infrastructure and possible multiple locations throughout the community provide for an ideal set-up of feeding sites or donation centers if necessary.

By being a support agency in ESF 6, libraries will then gain access to the Emergency Operations Center (EOC) where they will also be part of the larger group of ESF 6 agencies which usually includes Social Services, Mental Health, Volunteer Organizations, and Public Education. In aligning with these agencies libraries can provide input into a community's path to recovery.

Emergency Management in the United States has developed substantially since 1803 due to changes in law and public policy. The greatest changes have occurred in the past decade as a result of several substantial natural disasters. The acknowledgement that local preparation and response is far more effective than federal laws and national reaction. With this information responses, especially for localities that incorporate practices that are based on recommended mission areas and emergency support functions, are far more successful.

REFERENCES

Arnold, J. L. (1988). The Evolution of the Flood Control Act of 1936. *Office of History, United States Army Corp of Engineers*. Retrieved from http://www.publications.usace.army.mil/Portals/76/Publications/EngineerPamphlets/EP_870-1-29.pdf

Block, D. (2015, August 27). Lessons learned from Katrina, 10 years later. *Voice of America*. Retrieved from http:// www.voanews.com/a/lessons-learned-hurricane-katrina/2935755.html

Chesterfield County. VA. (August, 2016). *Chesterfield County, VA Emergency Management*. Retrieved from http://www.chesterfield.gov/em/

Federal Emergency Management Agency. (2015, April 22). *Core capabilities*. Retrieved from www.fema.gov/core-capabilities

Federal Emergenecy Management Agency. (2016, April). Pub 1. *Media Library; Federal Emergenecy Management Agency*. Retrieved https://www.fema.gov/media-library-data/1462196227387-c10c40e585223d22e2595001e50f1e5c/Pub1_04-07.pdf

Flynn, S. (2007). *The Edge of Disaster*. New York, NY: Random House.

Holdeman, E. (2010, January 4). A Decade of Disasters 2000-2009. *Emergency Management*. Retrieved from http://www.govtech.com/em/disaster/A-Decade-of-Disasters-2000-2009.html

Mabe, M. R. (2016). Libraries to the Rescue. In E. N. Decker & J. A. Townes (Eds.), *Handbook of Research on Disaster Management and Contingency Planning in Modern Libraries* (pp. 494–516). Hershey, PA: IGI Global. doi:10.4018/978-1-4666-8624-3.ch021

Sylves, R. T. (2015). *Disaster Policy and Politics* (2nd ed., pp. 59–72). Los Angeles, CA: CQ Press.

The Minnesota Governors Council on Developmental Disabilities. (n. d.). *Individuals with Autism Spectrum Disorder and Emergency Planning Resources*. Retrieved from http://mn.gov/mnddc/emergency-planning/la-ep-lawsuit.html

Wikipedia. (2016, May 17) *List of natural disasters in the United States*. Retrieved from http://en.wikipedia.org/wiki/List_of_natural_disasters_in_the_United_States

Chapter 2
The CCPL Model

As discussed, emergency management has changed over the years due to shifts in political approach, legal definitions, and the changing nature of disasters themselves. Dangerous events happen regularly in every community. Many public libraries have experienced a dangerous event becoming a disaster first hand. In the case of events that displace the population; e.g. hurricanes, floods, fires, earthquakes, libraries have become the serendipitous gathering place for survivors, victims, and responders. A variety of papers and reports have documented the types of responses public libraries mount simply because the community knows that the library exists and the librarians are willing to render service.

According to Zach (2011), a national survey in 2009 that yielded over 1500 responses from librarians who had recent experience responding to local emergency management situations, confirmed that librarians can make valuable contributions to emergency planning and response activities. Unfortunately, most libraries surveyed see their primary role during times of crisis as protecting library collections and maintaining normal services. An exploratory study by Hahn in libraries in Maryland and Alabama found that although governmental officials in these states could envision potential roles for libraries in emergency situations, libraries were rarely considered when emergency planning took place or included in local emergency management plans. A lack of seeing their mission beyond the normal service model during times of crisis combined with exclusion from preplanning and actual crisis response discussions reduces or eliminates the ability of the library to play a key role during local emergencies and disasters.

DOI: 10.4018/978-1-5225-2196-9.ch002

Copyright ©2017, IGI Global. Copying or distributing in print or electronic forms without written permission of IGI Global is prohibited.

In the case of Chesterfield County Virginia, the way events and circumstances unfolded in a positive manner since 2002 is a study in and of itself. Through several natural disasters library staff, emergency management professionals and county leaders all converged on the same path enough times that a more serious look at how library facilities and staff could be better utilized during disasters was warranted.

Their experience is supported by primary recommendations voiced by other library professionals and replicated by scores of other libraries in similar circumstances. The unique elements of their emerging model practiced locally includes the close partnerships maintained by the library and the emergency coordinator, the level of training library staff has received, and the numbers of times that the responsibilities expected of the library during a disaster have been activated.

There is evidence that a number of U.S. public libraries have collaborated with local emergency response agencies to provide services during crisis, including staffing emergency information centers, and providing access to the internet. However, little is known about how public libraries may be proactively engaging in community-wide disaster planning (Hagar, 2014). In the case of CCPL, their collaboration was deliberate and consistent over the years and included bouts of serendipity.

CCPL's first emergency management role as defined in the Chesterfield County Emergency Operations Plan before 2005 was to provide volunteers to staff the Emergency Operations Center public phone bank. Although not a part of the basic county plan, CCPL also provided a range of services before and immediately after natural disasters at its branch libraries. Unfortunately, the emergency services at the branch libraries had typically gone unnoticed because CCPL was not included the Incident Command Process where emergency needs and available resources were discussed and assignments made.

During Hurricane Isabel in 2003 CCPL provided services to 15,000 residents each day for several days after the hurricane subsided. These services included, internet access, charging stations, cooling stations, and obvious access to public restrooms and access to public data via local media outlets being broadcast in the buildings. During Tropical Storm Gaston in 2004 they remained open in the hours up to the storm and resumed operations immediately after the storm subsided again providing similar services to thousands. Acknowledging that the community used the library, beginning in 2005 county leaders included CCPL in the revision of the county's basic emergency plan. At that time, CCPL was assigned the responsibility of coordinating donations and volunteers during disasters situations. This action expanded CCPL's role and allowed them to engage in the Incident Command process during emergency situations. (Mabe, 2016)

For the next several years there was relative calm in weather and emergency crisis. CCPL continued to participate in the annual review of the basic plan and developed internal procedures to coordinate volunteers and donations during the next crisis. In 2011 Hurricane Irene turned out to be more powerful than expected. With a large portion of the county without power concerns with how and where to activating mass care functions were discussed. Unlike previous disasters where CCPL provided service outside the basic plan and EOC, during Hurricane Irene CCPL was part of the Incident Command process. During the mass care discussion in the Incident Command Process CCPL administration reported that customers were flocking to the as expected. After some logistical discussion the emergency management director recommended that they included a branch library in an area particularly hit hard by the storm as a possible location for mass care feeding location. Within 24 hours Red Cross were providing emergency meals at the LaPrade Branch library. Eventually 2000 meals were distributed at the library over three days. Customers were commented that the library had become their lifeboat (Mabe, 2016).

Due to their success, emergency staff and library administration discussed the need for additional training for library staff including providing Community Emergency Response Team (CERT) training. This training was provided to key library staff in January 2012 and prepared staff for the next event which happened six months later in early July 2012. In late June a freak storm hit the area knocking out power for short periods of time throughout the county for several days. A week later the EOC command staff activated mass car resources including the library to operate a daytime shelter during the Fourth of July holiday weekend. CERT certified library staff opened the main facility for hundreds of visitors during this period even though all other libraries and county facilities were closed for the holiday. CCPL's service during Irene and the 2012 freak event encouraged county administration to list CCPL in the basic plan as a temporary shelter and feeding station in future emergency situations (Mabe, 2016).

CCPL's example from 2003 through 2012 demonstrates a deliberate approach to work with local emergency management professionals to determine and define their contribution during local emergency situations. Four phases in their experience can be defined as their key to establishing a collaborative partnership between CCPL county emergency management similar to the many other local municipal agencies partnerships who also support local emergency management efforts.

ENGAGE IN THE LOCAL EMERGENCY RESPONSE PLANNING PROCESS

As previously noted, there is evidence that a number of U.S. public libraries have collaborated with local emergency response agencies. In the case of CCPL, the collaboration was deliberate and consistent on the part of CCPL. Early experiences indicate that like other public libraries, CCPL offered valuable services to customers locally during several natural disasters because customers showed up at the library seeking basic needs. Unfortunately, these services were not known to the emergency management leaders. In the case of the 2003 Hurricane Irene event, CCPL services went entirely unnoticed by emergency management and county officials because they were not included in the basic response plan, the Emergency Operations Center (EOC) or the Incident Command (IC) emergency reaction process where needs and resources are discussed and aligned for resolution. CCPL reasoned that regardless of their internal decision, customers were going to show up at the local libraries following a crisis. Being included in the local planning process would strengthen their ability to respond.

PARTICIPATE IN THE EOC COMMAND PROCESS

A plethora of disaster recovery services were identified in libraries along the Gulf Cost in 2004-2005. Library services included: responding to information inquiries, creating community contact centers, staffing shelters in library buildings, housing city command centers (i.e. police, fire, public works), distributing food and supplies, providing hook-ups to recharge electronics and communication devices, assisting with completion of FEMA insurance and other paperwork, providing library materials to evacuee shelters, providing FEMA, Red Cross National Guard and Army Corp of Engineers personnel with a place to meet with residents and giving temporary library cards to relief workers among other services (Veil & Bishop, 2012).

CCPL deliberate engagement with local authorities beginning in 2003 to discuss their contributions which mirrored those defined by other libraries in the Gulf Coast area is a contrast to the libraries that engaged the community independent of local emergency management plans and practices. Once emergency management staff were aware of CCPL's libraries evidence based services, resources and capacity, they willingly included them in the local emergency planning process. This action not only provided CCPL opportunity to officially coordinate their emergency services with other local agencies and departments who routinely respond during a crisis; but it also provided CCPL a literal seat at the table and a voice in the EOC and IC process where emergency reaction process discussions happen during the crisis.

There is significant support for the development of a wide variety of types of responses provided by information professionals during and after natural or accidental disasters. It appears that, while information professionals provide timely and relevant services in an appropriate format to meet the needs of their users, they often do not leverage their unique skills because they have not had the appropriate preparation for what to do in case of a disaster beyond that focused on preserving the institution itself (Zach & McKnight, 2010).

Examples of a few libraries working with local emergency management response agencies to provided essential information services during crisis do exist. For example, when terrorists flew a plane into the Pentagon in 2001, Arlington County (Virginia) Public Libraries served their community with critical information including twice daily police and fire department briefings, local traffic updates, special postings to firefighters and lists of volunteers. In 2003 as Hurricane Isabel hovered off the east coast, the Virginia Beach (Virginia) public library staff engaged with a community partner to staff telephones at the Emergency Public Information Center (Zach, 2011).

Like these libraries, CCPL recognized that they could not close the door and wait for the storm to blow out to sea. Customers intentionally flock to the library during times of crisis expecting the facility to be open so they can receive help and/or access to basic services and information. CCPL's involvement in the planning process and having a voice during reaction discussions allowed them to control their emergency crisis destiny. In time CCPL leaders realized that training staff to know what to expect during a natural disaster and how to respond strengthened the system's capacity to leverage the unique skills CCPL libraries and librarians could provide the community during the crisis.

CCPL has found that Citizen Emergency Response Team (CERT) training improves the library response and enhances staff confidence. CERT training provides valuable hands on experience in critical areas that prepare the individual to respond quickly as individuals and efficiently as a team. CCPL CERT trained staff are the first line of defense for the system and add a level of confidence to the system's ability to volunteer for duty when the moment happens.

FOLLOW THROUGH WITH THE ASSIGNMENT

No doubt that CCPL's ability to provide services to 15,000 residents a day for several days after Hurricane Irene in 2003 and remaining open in the hours leading up to Tropical Storm Gaston in 2004 and providing services to thousands immediately after demonstrated the value of CCPL in the emergency management process to local emergency management leaders. CCPL's reputation was enhanced years later when they coordinated with the Red Cross and distributed 2000 meals to needy residents after Hurricane Isabel subsided and provided a place out of the heat during a major summer holiday weekend in 2012.

Being able to perform and follow through on assignments is important in the emergency response process. Leaders assigned to manage during a crisis assess situations and assign resources to solve problems and save lives. Aside from being part of the written plan, participating in the EOC, and volunteering when required, follow through demonstrates to emergency leaders they can rely on your contributions

History does repeat. Emergency management continues to evolve. The professionals who manage during local disaster are always looking for critical partners interested in supplementing their meager resources and libraries are already included. CCPL's model suits their circumstances.

COLLABORATIVE PARTNERS

It is important to note that a "disaster" always requires outside resources that are normally allocated to public safety. Because disasters do happen, emergency management most commonly lives in a preparedness state of public education, training, and plan development. According to the FEMA (2014) a disaster is "a dangerous event that causes significant human and economic loss and demands a crisis response beyond the scope of local and state resource. Disasters are distinguished from emergencies by the greater level of response required." In the classic model of disaster response emergency management professionals by necessity must lean on a network of volunteers to meet most local mass care needs.

To ensure consistency in volunteer services, emergency management professionals rely on groups to provide most of the manpower in a response. In nearly every disaster these volunteer groups are delayed in response because of a lack of training and orientation in order to be staged and deployed into a disaster area. Volunteers that are known to be trained or are vetted ahead of time according to established standards are the most useful. The Volunteer Organization Active in Disaster (VOAD) (2016) recommends four fundamental values expected of serious emergency response volunteers that are classified by the four Cs of disaster preparedness response, relief, and recovery. These four Cs are Cooperation, Communication, Coordination and Collaboration. These values allow for non-profit and faith based groups to organize volunteers who descend on disaster areas to offer help and assistance confidently and effectively.

SERENDIPITY

One final area that needs to be addressed in preparation for a broader discussion is the frequency with which luck and circumstances play a role in a successful disaster response. Although disasters result in tragedy, responses frequently unfold in an unexpectedly positive manner.

David Campbell (2015) has been a technology executive, board member, investment banker, and humanitarian field volunteer during a large portion of his adult life. David's lifetime business background includes positions as president of BBN Technologies of Cambridge, and CEO at public companies Computer Task Group of Buffalo and Xpedior of Chicago.

David was on the verge of retirement when the massive tsunami hit Thailand in 2004. Armed with donations he collected from friends and 35 years of business management leadership, David traveled to Thailand determined to help. The experience was life-changing and led David to create and lead the nonprofit natural disaster response and rebuild organization, All Hands Volunteers.

Perhaps his business experience with strategic planning that lead him to coin the motto, Maximum Impact, Minimum Bureaucracy, for the All Hands Volunteers group that he respectively leads. His organization enables volunteers to provide hands-on assistance to survivors of natural disasters in the U.S. and abroad. Over the last 11 years, David has volunteered on dozens of post-disaster projects including Hurricanes Katrina and Sandy, earthquakes in Haiti, The Philippines, and Nepal, and flooding in Iowa and South Carolina. His organization has earned the highest possible 4 Star Rating by Charity Navigator and a top 5 Star rating from Greatnonprofits.org and GoOverseas.com.

David recognizes every disaster is different but is always surprised at how often serendipity shapes their strategy and contribution for success. Through the years, his experience has encouraged him to always look for the serendipitous moment in each response. For example, when the March 11, 2011 earthquake, tsunami, nuclear power plant disaster in Japan happened, David immediately made arrangements to go to Japan. Once there, making arrangements to get to an affected area was very difficult. The area was washed out, the roads buckled and the power plant was potentially toxic making it difficult to find any kind of public transportation to the area. Not speaking Japanese made it near impossible. As luck or serendipity, would have it, a Japanese insurance adjuster named Henry, who studied banking in the United States at Tufts University was assigned to help David. Within a few hours, his job as an insurance adjuster and language ability allowed him access to necessary government office and authorities to get David on a plane to the affected area the very next day. David has often wondered, what would have happened if Henry hadn't been assigned to him to help.

In 2006, Deborah Bunker (2006) at the University of Sydney Business School in Sydney, Australia was engaged in evaluating the use of information systems in disaster management. She found that while those in command and control use official information to manage disaster response and recovery it is not always the optimal approach. Her discovery was that the serendipitous use and generation of information by first responders, emergency workers, and the general public can be more useful to those caught up in the disaster than the information that comes from official channels.

Two separate instances of serendipity she references include the 2010 Haiti Earthquake and 2011 Libyan Civil Uprising. In each of these situations "crowd sourcing technologies" to spread real time information from those involved in the incident and/or encourage groups of volunteers to self-organize into ad-hoc groups to help organize and route information during the immediate aftermath of the incidents made the flow of information during these particular disasters much easier than expected, considering the circumstances.

Ms. Bunkers analysis reinforced "...the notion that individuals and groups do self-organize and that emerging patterns of behavior and technology are quickly learned from, noted and reused by human beings to respond to their immediate environmental problems" (Bunker, 2006).

As luck would have it, Chesterfield County Public Library (CCPL) with 10 branches scattered across a 421 square mile county has had only one branch library negatively impacted during several natural disasters over the past 13 years. While there can be no guarantees that this positive track record will continue, CCPL has begun to assume that all library facilities would be available in emergency situations.

This assumption allows the county to plan on all branches being made available to provide day time temporary shelter, through the duration of the disaster and during recovery as needed. And nighttime shelter being made available at the Central Branch Library, which is centrally located in the county and has a generously sized meeting room with independent entrances and exits and immediate access to rest rooms. The meeting room can also be isolated from the main public portion of the library.

Because the library facilities seem to always be available, CCPL and the county now share a current understanding that mass care feeding programs can be implemented at three branches located in the center of populations in the northern, central, and southern areas of the county. This plan assumes that customers are able to transport themselves to one of these locations. These services would also include emergency services personnel directing vehicle and pedestrian traffic in an efficient manner that will result in minimal impact on regular library business.

In chapter three, we will examine profiles of several public libraries that appear to have experienced serendipity in mounting a response to a local disaster. While one shouldn't assume serendipity, it has occurred often enough in previous disasters that it is worthwhile to have plans in place that will encourage public libraries to take advantage of the those lucky moments.

REFERENCES

Bunker, D. (2006). Serendipity in Disaster and Complex Scenarios. *Proceedings of the 1st International Workshop on Encouraging Serendipity in Interactive Systems, 13th IFIP TC13 Conference on Human-Computer Interaction,* Sydney (pp. 7-10). Retrieved from http://interact2011.org/

Campbell, D. (2015). *All Hands: The Evolution of a Volunteer-Power Disaster Response Organization.* Austin, TX: River Grove Books.

Federal Emergency Management Agency. (2014). *Independent Study 230 (IS-230): Principles of Emergency Management (Preface).* Washington, DC: United States Printing Office.

Hagar, C. (2014). The US public library response to natural disasters: A whole community approach. *World Libraries, 21(1).* Retrieved from https://worldlibraries.dom.edu/index.php/worldlib/article/view/548/472Pub

Mabe, M. R. (2016). Libraries to the Rescue. In E. N. Decker & J. A. Townes (Eds.), *Handbook of Research on Disaster Management and Contingency Planning in Modern Libraries* (pp. 494–516). Hershey, PA: IGI Global. doi:10.4018/978-1-4666-8624-3.ch021

Veil, S. R., & Bishop, B. W. (2012). Community resilience and public libraries: Post crisis information and connectivity. *Natural Hazards Center*. Retrieved from https://hazards.colorado.edu/uploads/quick_report/veil_2012.pdf

Volunteer Organizations Active in Disaster (VOAD). (2016, August 15). International-Committee-Guidance-Final. *Volunteer Organizations Active in Disaster (VOAD)*. Retrieved from www.nvoad.org/International/

Zach, L. (2011). What do I do in an emergency? The role of public libraries in providing information during times of crisis. *Science & Technology Libraries*, *30*(4), 404–413. doi:10.1080/0194262X.2011.626341

Zach, L., & McKnight, M. (2010). Innovative services improvised during disasters: Evidence based education modules to prepare students and practitioners for shift in community information needs. *Journal of Education for Library and Information Science*, *51*(2), 76–85.

Chapter 3
The Natural Role of the Public Library

WHY LIBRARIES AND LIBRARIANS?

Extreme weather events, natural disasters and unexpected natural or man-made community emergencies pose distinct problems for public policy theorists, political leaders, and practicing managers deciding how to mount a response. Preparing for a potentially deadly and destructive unexpected event or a natural disaster is a daunting task. According to Comfort (2002), most emergency professionals tasked with mounting a response to a natural disaster typically consider these events rare occasions and prefer to address mitigation along the lines of acceptable risk instead of applying more direct preplanned tactics. Experience has taught us that when extreme events or natural disasters occur and public agencies fail to respond promptly, directly and efficiently, the consequences are more severe. One assumes that the emergency manager in charge incorporates resources when determining acceptable risk to ensure a positive outcome. One seldom considered critical resource is the public library. With its iconic facilities, educated staff, and digital resource access, the public library already operates as a personal command center for customers seeking data, services, and resources to survive on a normal day. During unexpected extreme events, the library is frequently the unsung hero for the majority of the community in crisis.

DOI: 10.4018/978-1-5225-2196-9.ch003

Copyright ©2017, IGI Global. Copying or distributing in print or electronic forms without written permission of IGI Global is prohibited.

Within the five emergency management mission areas, there are seventeen Emergency Support Functions (ESFs) that guide actions in a general response to a large emergency or disaster. The National Response Framework put out through FEMA, outlines the ESFs at the Federal level. The National Response Framework also serves as the guiding document for state and local emergency operations plans, ensuring that resource requests and information sharing moves through all levels of government. Among the seventeen ESFs, the most natural fit for libraries is Emergency Support Function #6: Mass Care. It is through this function that human service needs promote the delivery of service and the implementation of programs to assist individuals, households and families affected by an incident. In this scope, emergency assistance includes providing recharging stations, collecting and providing information on victims or property to family members, providing temporary child care, and assisting in expediting the processing of new benefits claims.

Librarians perform similar tasks on a daily basis; making it a natural choice for emergency management personnel to enlist their aid during disasters. Emergency management should engage library staff in a developing role or relationship because their skill set may not always be known to public safety officials. Short term recovery needs that are housed under mass care including feeding, information needs, filling out paperwork, and temporary shelter parallel human services with public safety in the role of first responders. The only difference is first responders are regularly trained for emergency and disaster situations while human services, including libraries, are willing to help, but have not been invited to collaborate. Despite not traditionally being a contributing planned partner, libraries have organically helped as disaster needs arose in their communities.

Libraries must present their case to emergency management to show the benefits that their building and frontline staff can provide in disaster recovery. This was the case of Chesterfield County Public Library (CCPL) during Hurricane Irene (Mabe, 2016). The library director convinced the personnel working the emergency operations center that the libraries were their perfect partners. This relationship continues to evolve, due to a strong culture of teamwork and priorities that are well defined and consistent with the overall mission of the county. Norfolk Public Safety Director Jim Redick (2016) has suggested that the role ambiguity is replaced with role clarity. As the role of local human services continues to evolve, emergency management must engage these resources, specifically libraries, to provide whole community preparedness, response, and recovery.

The Natural Role of the Public Library

Historically libraries archive information and documents. For the modern public librarian, this historical definition of the profession is as incomplete and antiquated as library books with paper cards and pockets. Public librarians troubleshoot all forms of technology every day. How do I print? Where is my print job? Why can't I access my email? What is my PIN, password, etc.? Can I print pictures off my phone? I need a form for court, where is it? Why did it print gibberish? Why can't I sign on to county Wi-Fi? Why won't the eBook download? Why does Adobe Flash not work? What is wrong with this mouse, keyboard, number lock, computer? These questions are just a small sampling of the everyday technology problems librarians must solve and then explain to customers.

According to Chesterfield County Public Library librarian Lori Purcell (personal communication, August 16, 2016), besides quickly finding the required information, librarians must be able to relay this information without making the person asking the question feel foolish or embarrassed. Since librarians teach classes on many different topics, including how to use a computer or 3D printer, write a resume, or research family history using the popular Ancestry database, learning and teaching are integral parts of the give and take of librarian-customer interactions. Public librarians' standard training includes mental health assistance, de-escalating patron quarrels, dealing with irate patrons, helping youth in crisis, and working with police and EMTs when patrons need emergency help. Tempers flare and tensions rise at a moment's notice. The librarian must deal with every person in an orderly fashion while connecting these people to the resources they need.

The roles of libraries and librarians have shifted from access work and information storage to a dynamic, fast-moving, ever changing, and sometimes challenging endeavor to find information and convey it in a way that can be utilized by the public. Remembering that the public comes in all shapes, sizes, socioeconomic and education levels makes librarians empathetic, responsive, and adaptive to all situations. Not only do librarians need to know that the forms are on the FEMA website, they must to be able to guide a patron who does not know how to type or use a mouse to fill out and submit the forms. Anyone can use Google; however, how many people have unlimited time to filter through millions of search results and choose the correct options? With the many skills required to successfully administer daily public library service, it is natural for librarians to assume responsibility in each phase of disaster planning, response, and mitigation. Many newspapers report that libraries happened by chance to be involved in short term recovery as well as in immediate response to disasters. If included in preparedness planning, libraries will be better able to fulfill the essential role that they can and should provide.

Libraries are key institutions to any community. While the core of local government may lie in one area of a municipality, the libraries are evenly dispersed as part of each community. Library facilities are generally geographically convenient, are of relative size, and maintain basic public amenities including, restrooms, internet access, electrical power, etc., that are all logistically critical to responders and survivors. Normal day-to-day library services include access to goods and services, so changing the product from providing materials and programs to delivery of information for medical needs, disaster updates, and immediate community weather reports is rather simple. And let's not forget that in difficult times, access to books and information as entertainment and distraction is still relevant. The library reflects the culture and population of its community. The residents know the staff and the variety of resources that the library provides. In turn, staff are dedicated to their facility as well as the people they serve. Since library staff already research and understand the demographics and idiosyncrasies of their unique communities, they are a natural fit in disaster planning. Currently there is not a standard model that includes librarians in disaster management planning. Libraries are already well established in a community and are able to be staffed and opened quickly so librarians can interview disaster victims for actual needs. This dependable mechanism has proven to enhance a community during the marked low moments that disasters cause.

For public libraries to provide support during emergency situations, it is essential to conduct frank, open discussions with local emergency management professionals before disaster strikes. These discussions need to include agreement on what services and resources the library can provide. In the words of the CCPL director, "What can we do? How can we help?"

According to interviews conducted by Brobst, Mandel, and McClure (2012), over 200 library managers and information officials have identified eight service roles that have been generally defined by other library leaders who have responded to disaster or emergency events. These roles which are self-explanatory include:

1. Be a safe-haven.
2. Offer normal service.
3. Coordinate disaster recovery.
4. Serve as a cultural organization center.
5. Operate as an information hub.
6. Be an evacuee resource.
7. Improvise.
8. Gather and store oral histories of the event.

Each of these roles is described below by libraries that have experienced the role first hand.

1. SAFE HAVEN: CONSIDER A LIVING ROOM

According to Executive Director of the Mississippi Library Commission Sharman Smith (Perlman, 2006), a public library represents a safe place for most people. They are generally well constructed buildings and their locations are known to most of the community. Couple this with fast internet access and it is no surprise that the library is the place many residents turn to during a local emergency.

This was the case at Ferguson Municipal Library in 2015. Scott Bonner was the director at Ferguson Municipal Library for about a month when the turmoil and riots hit Ferguson, MO (Library Journal, 2015). Sensing the fear and concern of the community, Bonner kept the libraries open throughout the trying time, conveying the importance of the public library as a safe place for all county residents. Most notable was the service to kids who were left without a school but had the library to fall back on. Soon teachers and volunteers were showing up to help. Charitable donations of cash and books followed; which was a windfall for the small, meager library that operates on less than half a million dollars a year.

During Tropical Storm Isabel (National Weather Service, 2004), most residents of Chesterfield County were without electricity for up to two weeks while phone service and even cell phone usage was disrupted.

Power outages were a significant effect of the storm. From emergency management situation reports as well as individual power company estimates, about 6 million customers lost power at some point during the storm across the states of North Carolina (700,000), Virginia (1,800,000), Maryland, Delaware, West Virginia (1,416,000), Pennsylvania (900,000), and adjoining states (approximately 800,000). Though the power companies anticipated the storm and planned for its effects, the scale of power outages due to fallen trees was far more than expected. Potomac Electric Power Company (PEPCO) Holdings Senior Communications Representative Robert Dobkin called Isabel the 'worst storm they have ever experienced' and attributed the massive power outages to the extremely wet antecedent ground conditions across the Mid-Atlantic States. The wet ground conditions enabled Isabel's winds to fell more trees across power lines than under normal ground moisture conditions. (NWS, 2004)

As soon as the downed trees were cleared and roads became passable, people wanted news of the storm and they desperately needed to contact family members. County libraries with electricity opened which allowed the public to bring their families to a safe place and use various resources.

2. NORMAL SERVICE: BUSINESS AS USUAL

Libraries in Hunterdon, New Jersey Libraries quickly resumed normal business in the wake of Hurricane Sandy (Hunterdon County Democrat, 2012). While there were more customers using the libraries, the services they offered were typical. Like the Hunterdon Libraries, the Fanwood Memorial Library in New Jersey was a beacon of light for local residents suffering from Hurricane Sandy. Residents from Fanwood and adjacent communities made their way to the library to gain access to a wide range of services including applying for college, doing homework, reviewing contracts, staying in touch with family, working, applying for work, contacting FEMA, answering personal and business email, exchanging tips on how to cope with the storm, and solving problems, as well as the comfort of being with other people. (Suburban News, 2012)

In a June 26, 2007 NN/LM Disaster Ready Initiative interview with Marty Thompson, director of the Robert M. Bird Health Sciences Library of the University of Oklahoma Thompson speaks about his experience after the 1995 bombing of Oklahoma City (Thompson, 2007). When questioned about the roles of libraries and librarians in disaster planning, response and recovery efforts, Thompson replied that their first role was to teach. He and Brad Robison organized the Bioterrorism 101 class which became a disaster planning class for staff and customers. In class, they present situations and elicit response and discussion from the students. One of Thompson's main points is the importance of the library to the community.

I would tell you that my first priority is to take care of the staff. Then work on reestablishing the library. This is a lesson I learned from Lee [Lee Brawner, former director of the Metropolitan Library System]: the library is an important constant for the community. Let them know that you are OK; and it will give them the confidence to move forward. (Thompson, 2007)

Whether the situation is violence, weather-related or bioterrorism, most library directors and staff's thoughts always return to the public they serve.

The New York Public Library suffered virtually no structural damage due to Hurricane Sandy (Bayliss, 2012). The Queens library also fared well, although four branches were severely damaged. Nine of the 60 branches of the Brooklyn Library were closed. Sandy also caused severe winter weather in West Virginia where days later only 14 of 175 public libraries were out of service. In all of these locations, public libraries and librarians seemed to weather the storm, making it possible to be open to the public and provide traditional and emergency services. In Rockaway, NY, the Queens library sent its Book Bus to help community members charge their phones and electronic devices, access recovery information, and file for assistance.

According to Dankowski (2015), Princeton Public Library director Tim Quinn learned how critical his disaster role was to local residents after Irene hit in 2011. Staff of PPL had 4,500 people stream through the doors of the library in the immediate aftermath of Irene, 2,000 more than the daily average. He was not surprised to see a line of customers waiting when he arrived at 11:00 a.m. on October 30 to open the library after Sandy had subsided.

3. DISASTER RECOVERY CENTER: A SINGLE HOUSING LOCATION FOR ALL AGENCIES PROVIDING BENEFITS

Disasters allow the community to experience a library beyond the traditional services.

As previously stated, after a disaster the first step in getting disaster assistance is applying to FEMA. While most pre-teens, teens and millennials are comfortable doing everything on their phones and laptops, other groups need more help. For example, most people know that FEMA is the government agency to obtain help after a disaster. Unfortunately, the application process requires more than a few clicks of the mouse or taps on a phone. For the members of the public that have never used a mouse, touchpad or a computer and cannot afford tablets or phones, this process is overwhelming and seemingly impossible. The situation involves getting the people to the resources and translating the resources into information that the public can consume. Librarians are trained to help patrons will limited skill sets or abilities.

The skills of the librarian and their experience in interacting with customers on complex issues have been observed and noted by emergency professionals over the past several years during many major disasters. In 2007 FEMA noted that immediately after a disaster they are always looking for locations to set up emergency centers where people can fill out forms and receive recovery information. The Massachusetts Board of Library Commissioners (MBLC) and FEMA piloted a project in 2007 to place libraries in the center of disaster recovery (Massachusetts Board of Library Commissioners, 2008). Librarians trained to help customers access and interpret information and sturdy library facilities assuming a place in the center of the community is another vital connection. MBLC and FEMA have not discovered anything new; but they have identified what is logical.

On NPR's Morning Edition (National Public Radio, 2013), August 12, 2013, librarian Barbara Byrne-Goldie noted that not only did local residents use the library after Hurricane Sandy; relief groups utilized the New Dorp branch of the New York Public Library. "We had groups of FEMA workers," Byrne-Goldie says. "We had groups of Red Cross workers using our facility as a gathering place, and also to print out information about streets and what house hadn't they knocked on the door of yet" (NPR, 2013).

4. CULTURAL ORGANIZATION CENTER: A LIAISON BETWEEN EMERGENCY MANAGEMENT AGENCIES AND CULTURAL ORGANIZATIONS

First Lady Michelle Obama said, "Libraries and museums don't just preserve and promote our cultural treasures, they also enrich and enlarge our lives. And that's really one of the most powerful things that you all do each and every day" (Institute for Museums and Library Services, 2016). This is also true during disasters.

National Heritage Responders (NHR) formed to help the overwhelming number of school, public, and academic libraries, archives, museums, and cultural centers that needed restoration work after a disaster. The NHR services previously offered by AIC-CERT are referred to as the "SWAT team for visual art" by the Wall Street Journal (Catton, 2012).

NHR began in part as a reaction to hurricanes Katrina and Rita, which struck New Orleans in the summer of 2005. Conservators possessed the much-needed expertise to save collections, but the administrative structure to coordinate the distribution of those skills was lacking. The program has since developed into a cohesive volunteer group of strongly committed conservators and other collections care professionals who monitor disasters as they occur across the nation and respond accordingly (American Institute for Conservation of Historic and Artistic Works, n.d.).

Volunteer conservators, librarians, archivists, curators, registrars, and collections care specialists comprise the highly trained National Heritage Responder team. A valuable asset to these teams, librarians are uniquely qualified to offer conservation and preservation advice and help because these areas of expertise relate to traditional library material service. As libraries acquire materials, it becomes the duty of the staff to evaluate, maintain and in some cases to protect the materials.

Libraries experience the same issues as other public and private buildings - weather related disasters, insects, fires, and broken pipes, which means most librarians have rescued their own collections a time or two. With institutional knowledge and practical experience, librarians are inherently suited to help the National Heritage Responders.

The three main services offed by NHR(American Institute for Conservation of Historic and Artistic Works, n.d.) include a hotline that dispenses free advice by phone, volunteer team deployments which assess damage and organize salvage operations within the community, and online and print resources for institution and individual recovery.

NHR responses include:

- Offering professional assistance and advice to individuals, artists, and institutions affected by Hurricane Sandy.
- Working with other groups, they helped to recover cultural and historic artifacts damaged by the 2010 Haitian earthquakes.
- Assessing August 2011 flood damage to the Ward County Historical Society in Minot, North Dakota, and making recommendations (n.d.).

5. INFORMATION HUB: A TOWN HALL OR COMMUNITY GATHERING PLACE

During Hurricane Irene, CCPL locations were obvious places for mass communication. With so many residents flocking to the libraries before, during, and after the hurricane, EOC staff elected to have daily briefing bulletins posted on the electronic announcement boards in each library.

A week after county residents began to resume normal life, CCPL began its assigned role of coordinating volunteers and donations by assembling work teams to clean up debris. Approximately three dozen locations throughout the county received clean up assistance. In the wrap-up EOC meeting, it was acknowledged that CCPL had played an important role in a variety of ESF areas including public communication and mass care.

The library, with its nine locations, expanded its role as an emergency responder. The buildings themselves became community centers where citizens could charge their phones and computers, connect to the internet, and continue to conduct personal, work, school and business tasks. In addition, the essential comforts of the facilities such as restrooms, HVAC and furnishings, provided visitors the ability to relax in comfortable surroundings. Over 44,000 people came through the doors of the library branches during five days after the storm. The parking lots were full, as were the tables and inside.

During Hurricane Sandy, New York Times writer Michael Kimmelman reported that relief hubs sprang up in various places including the Rockaway Beach Surf Club. Rockaway was one of the places hardest hit by the storm. The Rockaway Beach Surf Club is a party space. Following the hurricane, the club's owner posted this note on Facebook, "If you need anything, come; if you have anything, bring it." The response was hundreds and thousands bringing tools, equipment, and resources to charge phones, get news, and fellowship. The club became a heartbeat for a struggling neighborhood (Kimmelman, 2013).

The Red Hook Initiative Community Youth Center in Brooklyn became a similar gathering place (Kimmelman, 2013). This center had a kitchen, a common room and art on the walls. While serendipity seemed to play a big role in providing relief for survivors of Sandy, Eric Klinenberg, a sociologist at New York University, has asked if the places like this should be designed. As noted, many places serve as emergency centers but libraries, already de facto community centers, could be designed with the essential tools and resources required during emergencies. Mr. Kimmelman's point is that "If serendipity can't be planned, it can be planned for."

According to Dankowski (2015), Louisiana State Librarian Rebecca Hamilton reflected on the impact of Katrina on libraries by stating the obvious, "Even people who weren't regular library users knew they would get help at a library." Despite the fact that Katrina caused 35 percent of the libraries in Louisiana to close at one point, the overall number of visitors to libraries immediately following the storm was down only 1 percent. Hamilton recognized libraries were at the center of recovery. Unfortunately, traditional disaster planning in libraries prioritizes collections recovery over business continuity, sending the message that libraries are not essential services. The experiences of Katrina and later Hurricane Rita encouraged Hamilton to revamp the emergency and backup plans for her state library, which remains a part of the overall state plan. When Hurricanes Gustav and Ike showed up in 2008, her plans were put to the test. About 95 percent of Baton Rouge lost power, but the state library did not. "The state didn't necessarily get better prepared," says Hamilton, "but we did."

Not surprising, the East Baton Rouge Library (Dankowski, 2015) provided services for Hurricane Gustav survivors for weeks after the storm, including providing a facility for refuge, internet access, status updates about the storm, hotel information, power restoration updates, and information on how to access emergency resources and how to apply for FEMA resources.

The community of Silverado in the eastern hills of Orange County, CA, is one of four entities in the small downtown area (Oder, 2008). Wildfires in 2008 dislodged the town's 1,200 households. This branch of the Orange County Public Library played an important role in supporting the dislodged residents temporarily living in a tent city in the region. The residents quickly became frustrated with the lack of information about the wildfires and their community. Staff at the Silverado Library spent two weeks collecting and relaying information between the command post at a nearby regional park and the tent city. This library is seen on most days as the clearing house for everything. Being the eyes and ears for the tent city was business as usual.

During NPR's Morning Edition (2013) feature "For Disaster Preparedness: Pack a Library Card?" August 12, 2013, Barbara Byrne-Goldie, librarian at New Dorp branch of the New York Public Library in Staten Island noted that the library was not damaged but neighbors were dealing with up to 16 feet of water and no basic services after Hurricane Sandy. "We even had people asking if they could use the restrooms to clean up a little bit," says Barbara Byrne-Goldie. "They still didn't have running water, or hot water. So we came in very handy as community centers, that's for sure."

6. EVACUEE RESOURCE: A FOCAL POINT FOR EVACUEES SEEKING INFORMATION AND INSIGHT

There are historic disasters, however, which require people to evacuate their area no matter how well prepared they are. In cases like these, such as Hurricane Katrina when residents were evacuated to Houston, the new local libraries were able to provide evacuees a place to receive information about their badly damaged community.

Chesterfield County Emergency Management Coordinator, Emily Ashley's years of experience suggest mass care needs are met through sheltering. Theoretically, people evacuate their homes, either because they are in the path of a storm or underprepared for a prolonged power outage and show up at the local building that is serving as a temporary shelter. They check in with very detailed information before being handed a sleeping mat and provided three meals a day. As the saying goes, a shelter is a lifeboat, not a cruise ship.

In some regards there is an element of surrendering, especially for those who have to take shelter for being underprepared. However, libraries have the physical infrastructure and staff not to act as a shelter but to invite the community to be active in their own recovery. Rather than handing residents a sleeping mat in exchange for personal information, a library that is open during a disaster implies the community is open for business and its residents should be too. Libraries have the ability to not only stay fully functional but can provide internet access and charging stations for cell phones as well as entertainment for children. This paradigm shift of a shelter versus a charging station implies the community is ready to recover quicker and expects everyone to play their part in the recovery, including those who were affected by the event. Additionally, there is no feeling of surrender. People are allowed to use the library services as much as they feel necessary.

In June of 2012, wildfires drove about 1,000 residents from their homes to wait out the fires in a temporary Red Cross evacuation center in Loveland, CO (Weiss, 2012) a community 10 miles north of Fort Collins. Staff from the Poudre River Library District in Fort Collins dispatched staff, laptops, and a projector to the Red Cross shelter in Loveland to provide information updates about the fire. Staff eventually expanded the initial service to include story times for restless kids. Library director Irene Romsa offered her staff and resources to the Red Cross. The Red Cross accepted the offer and considered them 'first responders' in the circumstances.

Oralia Garza de Cortes, an Austin, Texas children's literature consultant, and Lucia Gonzalez, Director of the North Miami Public Library, knew libraries had to help the 70,000 refugee children detained while crossing the southern United States border fleeing from Central America (Baron, 2016). These children, often traveling alone, are housed in shelters and detention centers with nothing to occupy their time. The initial, deceptively simple goal to deliver books and introduce library services became extremely difficult due to the complexities of dealing with the various governmental agencies. At the 2014 American Library Association conference, a taskforce was formed and the Children in Crisis program was created. Operating in shelters across several states, the initiative continues as the Children in Crisis team develops strategies to reach refugee children and their families.

Garza de Cortes notes (Baron, 2016) that each book carries a bilingual bookplate and insert donated by International Board on Books for Young People (IBBY) with instructions about how to get a library card. For many, that card becomes a lifeline. 'These kids are coming from places that don't have libraries. They have to navigate a very complicated process, and knowing they can go into a library for information, to use a computer, or get help finding other services, all for free, is critical,' says Garza de Cortes.

Similar to the Central American children and their families, Syrian refugees urgently need the public library and the information it provides. April Witteveen (2015) in her December 29, 2015, Library Journal article, Public Libraries Support Refugees features the Salt Lake City Public Library and the Louisville Free Public Library. Located in cities that serve as refugee relocation cities, these library systems have risen to the challenge to successfully welcome and educate refugees to the library's importance. Salt Lake City coordinates with the Salt Lake City Public Library to offer tours that connect refugees with the many library resources. After Brooke Young, manager of the Glendale Branch, presented library resources to community partner, University of Utah's University Neighborhood Partners Program, "our meeting room demand has tripled and we have had people [from the immigrant/refugee community] come in and use our computers for resume help and job hunting"(2015). Sophie Maier, who is an immigration services librarian with the Louisville Free Public Library, takes her show on the road. "I have my suitcases on wheels," Maier told Library Journal, "I set up outside Walmart." Maier also frequents workplaces that hire from the immigrant and refugee community, such as a local meat processing plant"(2015).

No matter what the crisis might be, information for evacuees, immigrants and refugees can be found at the public library. If necessary, the information and programs are delivered to groups needing assistance. By collaborating and building partnerships, libraries provide information and the all-important ability to retrieve this information.

7. IMPROVISE: DO WHATEVER IS NEEDED WITHIN RESOURCES

Directors and staff from a variety of library systems addressed issues, offered suggestions, and solved problems. Gloucester County, VA, Director of Libraries, Diane Rebertus noticed that during disasters her library staff was being sent home when they were perfectly capable of a response. As library staff became trained, their role in whole-community preparedness through disaster awareness is obvious. Through this example, the people of Gloucester County know that their local library is a place to get information during post-disaster response and recovery.

In New Jersey, Bernardsville Public Library's community room (Zavalick, 2012) became a beehive of activity after Hurricane Sandy. The hurricane took out power for several days in much of the borough including at the library. But with generators, the library was able to provide power to charge phones, and a place to spend the time waiting.

While some libraries provided obvious services such as Wi-Fi and electrical outlets as charging stations and Wi-Fi in New Haven, CT., other libraries provided programming as a form of distraction like the West Islip Library's (Piccoli, 2013) movies for kids. Perhaps the most notable service is clean restrooms for victims who need a place to clean up. According to Rose (National Public Radio, 2013), in the aftermath there is a need for a place to be when power is out, a need for programs on recovery, a constant demand for information, and sometimes a need to just talk.

For public libraries that have become players in the local mitigation process, addressing mass care concerns is a natural fit. In communities that have been affected by extreme events more than one time, many residents head to the library assuming relief will be available. According to Comfort (2002), when the county response includes the library and the librarian, the county has the advantage of reacting humanely. Including a human services type organization in disaster reaction ensures the majority of individuals interacting directly with the public during the emergency will provide information from a victim's perspective.

8. THE DEVELOPMENT OF ORAL HISTORIES FROM RESPONDERS AND SURVIVORS OF THE EVENT

After a disaster, people feel the need to share their experience. Oral histories can help survivors process the events and start healing. They also serve to educate those who were not present. Not only is the moment in time captured, the information is available for future generations, scholars, and government agencies. The material has many purposes: it can serve to champion the human spirit or community sorrow, as a reference of the disaster, and a plan for the future. Libraries, understanding the significance of oral histories, collect, and preserve these stories.

The Library of Congress (2009) added the interview collection "Surviving Katrina and Rita in Houston: A Survivor-Centered Storytelling and Documentation Project" directed by Carl Lindahl and Pat Jasper in conjunction with the American Folklife Center. This peer-to-peer collection is a digital oral history with over 400 interviews of Hurricane Katrina and Hurricane Rita survivors who relocated to Houston, Texas after the storms.

Similar to the Hurricanes Katrina and Rita oral history collection, the Brooklyn Public Library (2012) collected, recorded and documented Hurricane Sandy stories. These stories are a testament to hardship and the resiliency of the people in Brooklyn, New York.

Oral history is a powerful way of preserving and sharing the uniquely personal perspectives of individuals and groups who are witness to history. In the year since Hurricane Sandy hit this area, the Brooklyn Collection has recorded a number of these interviews, (some at the Central Branch of Brooklyn Public Library, others in the neighborhoods hardest hit), to archive and make available to the public for research. Through these stories we get to hear how people in our neighborhoods met an incredible force of nature, and persevered with the help of family and friends, and the community (2012).

For survivors and participants, their stories make up the fabric of a disaster. For others, this sharing of stories is a celebration of life and the inherent goodness of people. Libraries realize the importance of oral histories and go beyond the collecting of stories. They plan events to highlight and bring these stories to their public. On September 27, 2016, the Mid-Manhattan branch of the New York Public Library hosted Regina Ress who performed her story, Compassion, Generosity, and Grace: Stories from 9/11 (New York Public Library, 2016). In her blog Young Adult Librarian, Haydee Camacho, shares Ress' message of hope and love. This storytelling event demonstrates emotional devastation paired with human compassion. The underlying message is the goodness and community of the people of New York.

Beyond these examples describing the eight service roles, thousands more exist whether documented or not. Libraries spring into action after disasters hit their communities. There are many logical roles libraries can play in support of disaster response because they have facilities and provide professional information services. The ingenuity and resourcefulness of library staff allow for the unlimited interpretation of these eight service roles. While observation indicates many roles that the library can absorb, most roles are taken on because a library facility exists and librarians are available to provide a range of services, some planned and some unplanned, while still providing existing library services.

REFERENCES

American Institute for Conservation of Historic and Artistic Works. (n. d.). National Heritage Responders (NHR). Retrieved from http://www.conservation-us.org/emergencies/national-heritage-responders#.WIqAD1PR-1t

Baron, M. (2016, May 16). Children in crisis: Libraries respond. *slj.com*. Retrieved from http://www.slj.com/2016/05/industry-news/children-in-crisis-libraries-respond/#_

Bayliss, S. (2012, November 1). New York City libraries relatively unscathed: New Jersey still taking stock. *Library Journal*. Retrieved from http://lj.libraryjournal.com/2012/11/industry-news/new-york-city-libraries-relatively-unscathed-new-jersey-still-taking-stock/#_

Brobst, J. L., Mandel, L. H., & McClure, C. R. (2012). Public libraries in crisis management: Roles of public libraries in hurricane/disaster preparedness response. In C. E. Hagar (Ed.), *Crisis Information Management Communication and Technologies* (pp. 155–172). Oxford: Chandos Publishing. doi:10.1016/B978-1-84334-647-0.50009-6

Brooklyn Public Library. (2012). *Brooklyn collections: Brooklyn Public Library*. Retrieved from http://www.bklynlibrary.org/brooklyn-collection/brooklyn-collection-hurricane-sandy-oral-history-project

Catton, P. (2012, November 19). Art conservators take up the fight. *wsj.com*. Retrieved from http://www.wsj.com/articles/SB10001424127887323551004578123210560794482

Cloak, M. (2012, July 4). Life in the summer without power is no fun. *Richmond Times Dispatch*, p. 1A.

Comfort, L. (2002). Managing intergovernmental responses to terrorism and other extreme events. *Publius*, *32*(4), 29–29. doi:10.1093/oxfordjournals.pubjof.a004970

Dankowski, T. (2015, August 25). Business continuity over collections recovery: How Louisiana's state library changed the way it plans for disaster. *American Libraries Online Magazine*. Retrieved from https://americanlibrariesmagazine.org/blogs/the-scoop/business-continuity-over-collections-recovery/

Emergency, M. W. R. (2016, January 10). American Institute for Conservation-Collections Emergency Response Team (AIC-CERT) becomes National Heritage Responders (NHR). *Emergency Managers Weekly Report*. Retrieved from http://emweeklyrpt.wordpress.com/2016/01/10/american-institute-for-conservation-collections-emergency-response-team-aic-cert-becomes-national-heritage-responders-nhr/

Hazard, C. (2003, September 19). Many trees falling like dominoes. *Richmond Times Dispatch*. p. 7A.

Hunterdon County Democrat. (2012, November 1). *Hunterdon libraries take up slack after Hurricane Sandy victims lose home internet and wi-fi*. Retrieved from http://www.nj.com/hunterdon-county-democrat/index.ssf/2012/11/hunterdon_libraries_take_up_sl.html

Institute for Museums and Library Services. (2016, November). *IMLS at 20: IMLS history*. Retrieved from https://www.imls.gov/imlsat20

Kimmelman, M. (2013, October 2). Next time, libraries could be our shelters from the storm. *New York Times*. Retrieved from http://www.nytimes.com/2013/10/03/arts/design/next-time-libraries-could-be-our-shelters-from-the-storm.html

Library Journal. (2015, March 20). Scott Bonner / movers & shakers - community builders. *Library Journal*. Retrieved from http://lj.libraryjournal.com/2015/03/people/movers-shakers-2015/scott-bonner-movers-shakers-2015-community-builders/

Library of Congress. (2009, January). *American Folklife Center & Veterans History Project: Library of Congress Annual Report, Fiscal Year2009*. Retrieved from https://www.loc.gov/folklife/annualreport/AFCAnnualReport-FY2009wcbsm.pdf

Mabe, M. R. (2016). Libraries to the Rescue. In E. N. Decker & J. A. Townes (Eds.), *Handbook of Research on Disaster Management and Contingency Planning in Modern Libraries* (pp. 494–516). Hershey, PA: IGI Global. doi:10.4018/978-1-4666-8624-3.ch021

Massachusetts Libraries Board of Library Commissioners. (2008, November 07). Libraries to serve as disaster recovery centers. Retrieved from http://mblc.state.ma.us/mblc/news/releases/past-releases/2008/nr081107.php

National Public Radio. (2013, August 12). For disaster preparedness: Pack a library card? *Morning Edition, National Public Radio*. Retrieved from http://www.npr.org/2013/08/12/210541233/for-disasters-pack-a-first-aid-kit-bottled-water-and-a-library-card

National Weather Service. (2004, May 16). Hurricane Isabel NWS Service Assessment. United States Department of Commerce, *National Oceanic and Atmospheric Administration, National Weather Service*. Retrieved from https://repository.library.noaa.gov/view/noaa/6471/noaa_6471_DS1.pdf?

New York Public Library. (2016, September 22). Sharing stories with Regina Ress. *nypl.org*. Retrieved from https://www.nypl.org/blog/2016/09/22/sharing-stories-regina-ress

Oder, N. (2008, May30). When a fire hit, a rural OC library was there to help. *Library Journal.*

Perlman, E. (2006, December). Critical connectors. *Governing*, 57.

Piccoli, K. (2013, October 31). West Islip Public Library open, offering charging stations and wifi. *West Islip Patch.* Retrieved from http://patch.com/new-york/westislip/west-islip-public-library-open-offering-charging-stat38f80c3a84

Redick, J. (2016, March). Regional Disaster Planning: Observations and Recommendations, A Case Study of Hampton Roads. *Paper presented to Disaster Planners in the Hampton Roads Region*, VA.

Suburban News. (2012, November 8*). Fanwood Library shines as a community hub during Hurricane Sandy.* Retrieved from http://www.nj.com/suburbannews/index.ssf/2012/11/fanwood_library_shines_as_comm.html

Thompson, M. (2007, June 26) Online Interview with Marty Thompson, Director of the Robert M. Bird Health Sciences Library at the University of Oklahoma. *National Network of Libraries of Medicine*. Retrieved from http://web.archive.org/web/20090118212137/http://nnlm.gov/ep/2007/06/26/robert-m-bird-health-sciences-library-university-of-oklahoma/

Weiss, L. B. (2012, June 19). CO serves kids "on the fly" as wildfires rage. *slj.com.* Retrieved from www.slj.com/2012/06/public-libraries/co-library-serves-kids-on-the-fly-as-wildfires-rage/

Witteveen, A. (2015, December 29). Public libraries support refugees. *lj.com*. Retrieved from http://lj.libraryjournal.com/2015/12/public-services/public-libraries-support-refugees/#_

Zavalick, C. (2012, November 9). Bernardsville library serves as a multi-purpose center. *Bernardsville News*. Retrieved from http://www.newjerseyhills.com/bernardsville_news/news/bernardsville-library-serves-as-multi-purpose-center/article_a8faeaf8-29b6-11e2-bc59-001a4bcf887a.html

Chapter 4
The Local Command Structure and How the Library Fits In

Many library directors may not know where to begin to be more involved in disaster response. It is critical to form a partnership with emergency management during non-emergency times as this alliance allows for trust and a strong foundation to be explored and created. When a disaster strikes, the pace is frantic and new concepts are hard to conceive; however, in the planning phase new ideas are allowed consideration. A forum that would work well for this partnership is a locality's Local Emergency Planning Committee (LEPC). In 1986 Congress amended the Comprehensive Environmental Response, Compensation, and Liability Act to the Superfund Amendments and Reauthorization Act (SARA). From the United States Environmental Protection Agency's website https://www.epa.gov/superfund/superfund-amendments-and-reauthorization-act-sara, "SARA reflected EPA's experience in administering the complex Superfund program during its first six years and made several important changes and additions to the program" (U.S. Environmental Protection Agency, 2016).

Furthermore, according to the EPA (2016) in the late 1980s after the events of Three Mile Island and West Virginia hazardous materials disasters, Congress enacted Title III of the Superfund Amendments and Reauthorization Act (SARA Title III) as national legislation on community safety. Among

DOI: 10.4018/978-1-5225-2196-9.ch004

Copyright ©2017, IGI Global. Copying or distributing in print or electronic forms without written permission of IGI Global is prohibited.

other things, this Congressional Act required each state to appoint a State Emergency Response Commission (SERC) that would divide the state into emergency planning districts lead by a Local Emergency Planning Committee (LEPC).

The LEPC includes a mix of public safety, private sector facilities that have hazardous materials, utilities, public health, communication groups, and emergency management. For a majority of communities, the LEPC's mission grew in the aftermath of 9/11. January 2002, President George W. Bush issued a call to action for people to volunteer and capture the spirit of service that emerged in communities after the horrific attacks (www.read.gov/about-citizen-corps). This action led to the development of Citizen Corps which under the Department of Homeland Security is tasked with providing training to assist in the recovery after a disaster or terrorist attack. Citizen Corps is the parent funding source for Community Emergency Response Team (CERT) and through the Department of Homeland Security's competitive grant funds; localities are able to apply for Citizen Corps funding which includes CERT, Neighborhood Watch, and Medical Reserve Corps training. A need arose for an established body to supervise grant guidance and review spending and for many localities the LEPC fulfilled this role. With the focus not only on community involvement but also training, the LEPC is a robust committee of learning, best practices, and a variety of viewpoints most often led by Emergency Management. For libraries that want to be involved in Emergency Management they would be welcomed at their local LEPCs to begin a discussion of mutual post disaster benefit.

Interestingly, libraries became the obvious choice for archiving and making these plans available to the public because they were known locally and trusted to safeguard this type of information. This obscure fact suggests the possibility that libraries already fit within the official command structure due to possible inclusion in an existing LEPC.

While the LEPC offers an outlet to begin a conversation about public library response to local disasters, it is also crucial to understand the local government structure. Local government is frequently addressed during secondary school civics. For the general public, their knowledge of local government is cursory and often ignored until an undesirable outcome of government service is noticed. For the library to successfully engage in the local emergency command structure there must be a clear understanding of local municipal government operations and the key players in disaster response.

According to Rubin's Foundations of Library and Information Science (2010), most public libraries in America are supported by a combination of taxes from state and local jurisdictions and governed by a board of directors appointed by local officials to serve the public interests (p. 165). These factors tie responsibility for public library operations and oversight to the local municipal government structure including legally binding jurisdictional policies and procedures.

To be effective in any of their professional operations, a public library must have a fundamental understanding of how their local government operates. Similarly, to be effective in their developing role in responding to natural disasters and emergencies, libraries must also have a thorough understanding of their local government structure including how they are included in local emergency management practices. This chapter will provide a brief overview of the structure and operations typical of most local municipal governments including forms of governance, the various departments that make up the local emergency management organization, and a thorough review of the factors that drive local emergency management principles and practices. The role of the public library is also addressed.

According to the National League of Cities (n. d.), the Constitution of the United States reserves authority-giving powers to the states. State political power is divided into three spheres: the local government, the state government and the functions that the two governments share. At the local level, there are four categories of discretionary authority:

- Structural -- power to choose the form of government, charter and enact charter revisions.
- Functional -- power to exercise local self government in a broad or limited manner.
- Fiscal -- authority to determine revenue sources, set tax rates, borrow funds and other related financial activities.
- Personnel -- authority to set employment rules, remuneration rates, employment conditions and collective bargaining.

Typically, the broadest discretionary powers are applicable to local government structure, and the narrowest are given to finance. Also, local governments endowed with discretionary authority may not always exercise it; for example, the adoption or amendment of a local government's municipal charter is infrequent.

Narrow Government Authority: Dillon's Rule

Dillon's Rule states that a substate government may engage in an activity only if it is specifically sanctioned by the state government. State constitutions vary in the level of power they grant to local governments. However, Dillon's Rule states that if there is a reasonable doubt whether a power has been conferred to a local government, then the power has not been conferred. This rule allows a state legislature to control local government structure, methods of financing its activities, its procedures and the authority to understake functions. The U.S. Supreme Court upheld Dillon's Rule in 1903 and again in 1923. Since then, the following tenets have become a cornerstone of American municipal law and have been applied to municipal powers in most states:

- A municipal corporation can exercise only the powers explicitly granted to them.
- Those necessarily or fairly implied in or incident to the powers expressly granted.
- Those essential to the declared objects and purposes of the corporation, not simply convenient, but indispensable (National League of Cities, n. d.).

LOCAL MUNICIPAL GOVERNMENT STRUCTURE

Local municipal governments are defined by a variety of names according to size of the community ranging from larger cities such as New York City to smaller boroughs or villages such as Lost Springs, Wyoming. Larger geographic regions are known as counties and are used to gather and collect population statistics and frequently operate the local government. Most local government structures are defined by the state constitution which allocates authority to local municipalities to collect taxes and deliver public service. In communities outside these designated municipalities the state constitution allows for district entities to provide local government services (Whitehouse, 2016).

Local municipalities generally assume responsibility for providing libraries, parks and recreation services, police and fire departments, housing services, emergency medical services, municipal courts, transportation services and infrastructure or public works (streets, sewers, snow removal, signage). A municipal government's power, granted by the state, is given to mayors, city councils, and other governing bodies such as library boards. The various forms of government and elected councils may fall into five main forms or categories of government and are briefly defined below.

FORMS OF LOCAL GOVERNMENT

Regardless of community size or the nature of the disaster, local government leaders are responsible for overseeing all four phases of emergency management – preparedness, response, recovery, and mitigation. Federal and state governments play a supporting role in the immediate aftermath and in providing funding and guidance for long-term recovery and mitigation.

Preparation and response, half of the emergency management cycle, generally get the most attention, particularly in high-risk areas. Preparing to respond usually involves significant training and practice to ensure that key local employees and supporting resources are ready to jump into action quickly and that local residents understand their roles and responsibilities in preparing for and responding to disasters.

Local government leaders, particularly those who have been through a major community disaster, recognize that preparing for long-term disaster recovery demands as much attention as preparing for short-term response. After a major disaster, the recovery process takes months and even years to bring a community back to a "new normal" and as strong as or better than before the disaster (Becker, 2009).

According to International City/County Management Association (ICMA) (2016), there are five forms of local government that are generally representative of the geographic size of the municipality. These forms of local government provide an opportunity for elected officials, municipal staff, and voters to engage in representative governing and to ensure oversight of the local municipal government operations and departments. Understanding the municipal structure and its function is critical to the public library's daily existence and key to its ability to officially engage in a local emergency management response.

Council-Manager

An elected council is the governing body of the municipality. They are aided by a municipal manager, hired by the council who enforces council policies. The council usually consists of five to nine members. One of the elected council members is designated the president as defined in the city charter.

Mayor-Council

The mayor or elected executive is designated as the head of the city or county government. The extent of the manager's authority is very broad and can range from ceremonial functions to day-to-day operations and include: hiring and firing department heads, preparation and administration of the budget, and veto power over acts of the elected council. The council adopts the policies and the budget, passes resolutions, and performs audits.

Commission

A board of commissioners serves as the elected head of the government. They adopt the budget, pass resolutions, and enact ordinances and regulations. Other needed officials are elected to serve as heads of major county departments such as sheriff, treasurer, and clerk.

Town Meeting

All voters of the town gather on a given day to elect a board of officers, generally called selectmen, and to make policy decisions. The elected board of selectmen has the responsibility for carrying out the policy set by the citizens. Frequently a manager or administrator is appointed to manage the administrative operations of the town.

Representative Town Meeting

Representative town meeting operates like the town meeting form. In the town meeting selectman are elected, in this form many citizens are chosen to represent them by vote in all matters of town business. Every member of the town is encouraged to attend the meeting, but only those chosen as representatives have a direct vote.

Within any of these forms of government are local service departments managed by professional staff whose mission is to provide the community the designated service. Most states require local municipalities to maintain a local emergency services manager who accesses other department staff and resources capable of responding to a disaster.

LOCAL EMERGENCY MANAGEMENT DEPARTMENTS

Within each of these forms of municipal government a variety of municipal service departments are maintained by law by the elected officials that can respond in a disaster. These service departments range from education and public safety to public works and recreation. It is within this structure that local emergency management has emerged. To be fully effective a municipal emergency management program requires several components. These components begin with the legal authority to operate a program of disaster related activities (FEMA, n. d.).

Most counties and large cities, and many towns in the U.S. have a local emergency management program and program manager. The larger or more hazard-prone the community the more likely it will be that there will be a full-time, paid emergency manager. Smaller jurisdictions may only have a part-time emergency manager or a volunteer emergency manager (either part-time or full). In addition, the larger and more hazard-prone jurisdictions are more likely to have a local emergency management staff. In summary, local emergency management agencies can range from part-time volunteers with few resources and little authority to large, highly professional organizations with state-of-the-art information technology.

In Virginia, the duties of Emergency Management are carefully laid out in the Code of Virginia Title 44, Chapter 3.2, Sections 44-146.19 (Emergency Services and Disasters Law). The code not only requires the Emergency Operations Plan to be adopted by each locality every four years to reflect changes in Virginia regulations, operational capabilities, lessons learned, and best practices in emergency management but there is also a protocol that a Director and Coordinator of Emergency Management must be named in each locality. Furthermore, this code outlines how to declare a disaster, notification procedures and a preliminary outline of disaster services provided by the Stafford Act. This code is the primary foundation block for any local emergency management program in Virginia (Code of Virginia § 44-146.19). While there are other laws such as SARA Title III and the Disaster Mitigation Act of 2000 that require plans and committees to be carried out they do not require the extensive program oversight that the Emergency Services and Disaster Law has.

Where there is an emergency manager that person is generally appointed, at the county level, by a county council or by county commissioners, depending on the form or structure of the county government. At the town or city level, the responsible emergency management official might report directly to a mayor or mayor's assistant, to a city or town manager, or to a department head, as in fire or public safety.

As noted previously, in most states emergency management is outlined in the state code. Frequently the code mandates that local governments have both a director of emergency management and emergency management coordinator. Meeting this requirement allows the municipality to receive Federal Emergency Management Performance Grant Funding to help fund these positions.

In the U.S. reliance, at all levels of government, for disaster response is placed on existing emergency services. Thus emergency service and response organizations have as their primary responsibility the response to day-today emergency and disaster situations. Generally these agencies are part of the traditional units of local government—it's one of the reasons local governments exist. The primary local emergency response organizations usually are (FEMA, n.d.):

Fire Departments

These can be volunteer, full-time paid, or a combination of both. In many small jurisdictions citizens still band together to train and to fight fires. The trend, however, is towards paid and around-the-clock on-call departments. Many, if not most, fire departments are increasingly becoming involved in areas other than fire, such as hazardous materials response, emergency medical assistance, and search and rescue operations.

The fire services are a front-line and critical emergency response group. Few other professions assume on a daily basis as much responsibility for disaster and emergency response.

Law Enforcement

Police and sheriff departments are the traditional local law enforcement agencies in local US jurisdictions. Their primary function is to maintain law and order and provide protection to the residents of a community. During an emergency, they are called upon to maintain law and order. Traffic control and scene security are frequent disaster responsibilities. They are also called upon to maintain the safety of emergency response personnel, more frequently from the thoughtlessness of onlookers rather than from overt hostile acts. Police are also often perceived by the public as a point of contact for information in an emergency or disaster.

Emergency Medical Services (EMS)

Emergency Medical Services are traditional responders in emergencies where casualties or injuries are involved. EMS personnel provide emergency medical treatment in the field and arrange or provide transport to hospitals when necessary. EMS services may be organized as an independent agency, a part of the Fire Department, a contracted or privately provided service, or a hospital based service.

Finance Department

Finance Departments are responsible for emergency purchasing and record keeping (important when reimbursement is sought from Federal or State agencies).

Municipal Utilities

Municipal Utilities are responsible for the restoration of damaged utilities.

Public Works Department

Public Works Departments typically are involved in debris removal and clean-up after a disaster—particularly of public property and roadways. Also much involved in pre-disaster mitigation efforts such as sand-bagging, grading and other engineering activities.

Office of Human Resources

An office of Human Resources can be responsible for the coordination of volunteers and the processing of temporary employees hired for disaster-related duty.

Building Inspection Departments

Building Inspection Departments are frequently involved in or responsible for damage assessment and condemnation of damaged buildings. Can also be involved in the screening of contractors, or the provision of guidelines on choosing reputable contractors (to repair hazard damaged property).

Non-Governmental Not-For-Profit Agencies

There are a number of non-governmental organizations throughout the US that have developed emergency or disaster response capabilities. Foremost among these is the American Red Cross, which is organized in chapters in most counties in the country. They provide assistance both to emergency or disaster victims and to emergency response personnel. Further information on the role of volunteer organizations active in disaster will be provided in a subsequent unit (FEMA, n.d.).

Public Library. Like most of these other municipal departments, the public library is an agency of local government charged with providing access to information to residents. According to International Federation of Library Associations and Institutions (IFLA) (2001), a public library is an organization established, supported and funded by the community, either through local, regional or national government or through some other form of community organization. It provides access to knowledge, information, and works of the imagination through a range of resources and services and is equally available to all members of the community regardless of race, nationality, age, gender, religion, language, disability, economic and employment status, and educational attainment. The emphasis on providing access through a range of services includes the types of services needed during emergency management situations.

Faith Based Organizations. These organizations play a large role in disaster recovery. These groups are usually established as part of a Disaster Recovery Task Force led by Emergency Management in a locality. These groups meet regularly to identify service needs and avoid duplication.

Emergency Management. This department has contact with these groups during non-disaster times by participating through the Volunteer Organization Active in Disasters (VOAD), which usually has an active chapter in each state. Furthermore, volunteer organizations are also reflected in emergency operations plans that allow for a clear request of what service is needed and how/when officials can assign the agency to fulfill requests.

Social Services. While not on the front line of defense like fire and law enforcement, social services in most locations is mandated to provide shelter. Social Services must be trained in sheltering management and offer support to those who have been evacuated from a high-risk area in anticipation of an emergency or in response to an actual emergency. Social services are also critical in the response phases of the disaster. Vulnerable populations struggle daily and a disaster puts their issues under a microscope. Social services can assist with these vulnerabilities by offering feeding services, temporary shelter, or supplemental benefits.

LOCAL EMERGENCY OPERATION FOUNDATION PRINCIPLES AND PRACTICES

Since the events of 9/11 and Hurricane Katrina in 2005, the local approach to natural disasters has changed significantly. Hurricane Katrina is known as one of the deadliest and most devastating natural disasters to occur in the history of the United States. With Katrina fresh on everyone's mind, Dr. Wayne Blanchard (2007) of FEMA's Emergency Management Higher Education Project convened a group of emergency management practitioners and academics to define what is emergency management. Blanchard and his group identified several relevant, non-civil defense oriented principles based on what worked and didn't work in modern emergency management practices. They recommended that these principles be used to develop a more effective doctrine of emergency management (Blanchard, 2007).

- Comprehensive – emergency managers consider and take into account all hazards, all phases, all stakeholders and all impacts relevant to disasters.
- Progressive – emergency managers anticipate future disasters and take preventive and preparatory measures to build disaster-resistant and disaster-resilient communities.
- Risk-driven – emergency managers use sound risk management principles (hazard identification, risk analysis, and impact analysis) in assigning priorities and resources.
- Integrated – emergency managers ensure unity of effort among all levels of government and all elements of a community.
- Collaborative – emergency managers create and sustain broad and sincere relationships among individuals and organizations to encourage trust, advocate a team atmosphere, build consensus, and facilitate communication.
- Coordinated – emergency managers synchronize the activities of all relevant stakeholders to achieve a common purpose.
- Flexible – emergency managers use creative and innovative approaches in solving disaster challenges.
- Professional – emergency managers value a science and knowledge-based approach based on education, training, experience, ethical practice, public stewardship and continuous improvement (Blanchard, 2007).

While the principles stemming from Dr. Blanchard's research are making the management of natural disasters easier to face, one fact cannot be forgotten; natural disasters cannot be thwarted or contained. On the other hand, plans can be developed that consider the full impact on victims. The guiding principles listed above have been helpful among the professionals who respond to natural disasters in all areas. And while emergency management approaches themselves cannot per Blanchard (2007) prevent disasters, the risk of injury and loss of life can be mitigated through the implementation of effective planning.

Blanchard (2007) notes that effective plans cover the four basic areas of prevention, preparedness, response, and recovery. These four basic areas tie back to the earlier established mission areas that FEMA outlines. The United Nations International Strategy for Disaster Reduction (2005) recommends the incorporation of the Hyogo Framework. This international initiative offers guiding principles, priorities for action, and practical means for achieving disaster resilience for vulnerable communities. Communities that develop long-term prevention plans that include evacuation strategies and environmental/design standards can mitigate the impact of disasters.

Mitigation is the most effective means of reducing the impact of disasters. Mitigation physically removes the threat from occurring such as removing a structure out of a floodplain or significantly reduces a threat such as home elevation. However, not all mitigation practices provide for a positive cost-benefit and cannot be completed. In these cases, an equally effective approach to reducing disaster impact is through emergency preparedness.

Community-based preparedness and management plans should be a high priority in local emergency management practice. Professionally developed plans can help minimize damage and loss of life by focusing on relief activities including rescue, relocation, disease prevention, providing food, water, temporary shelter, and emergency health care. These actions will go a long way in helping the victims of disaster.

Preparedness programs such as FEMA's Make a Plan, Get a Kit, Stay Informed or Ready Virginia allows residents to locate information in one place for all of their needs. Having this information enables them to research and stay engaged. Communities can also offer classes that provide first aid training, supply residents with a preparedness kit, and give information on how to make a personal emergency plan.

Plans should also address actions to be taken after a crisis has eased by incorporating recovery measures essential to addressing the short and long term vulnerabilities. Rebuilding the infrastructure, restoring telecommunications and transportation networks, and providing health care and rehabilitation services will all need to be addressed in the aftermath of a disaster. Pre and post response activities should blend with the development of other initiatives, such as training response staff and implementing policies and practices to avoid similar situations in the future. FEMA has a National Response Framework to outline what recovery looks like, localities can also look to their Continuity of Operations Plans or their Comprehensive Plan on how to restore services or rebuild a community from catastrophic disaster.

HOW DOES THE LIBRARY FIT?

There does not appear to be any emergency management protocol for how a library responds during a disaster or what its role may be. Eyewitness reports and observations indicate most libraries demonstrate that they can respond even without the necessary partnerships established beforehand.

In 2001, Harrington suggested six facility development trends that would be incorporated into future library facilities. They include self-service and operations efficiency, extreme flexibility, sustainable design, schools/libraries collaboration, aesthetics, and local customization. Many of the trends were based on the need for flexibility for staff and customers (Harrington, 2001).

Library historian Wayne A. Wiegand (2011) published a book that documented the communities' use of the library space in four locations between the years 1876 – 1956. The value of the public library as a public space for civic events and meetings of local groups is shown in the history of the four libraries in Wiegand's book. He stated that these public libraries were not established to further democracy. There were no clear references to the library having challenged local thinking or informing the resolution of controversial political issues. Their primary purpose and mission, as crafted over the generations by local leaders and users, was to foster the kinds of social harmony that community spaces and shared and experienced stories provide (p.186). They were successful to the degree that they were rooted in the community, engaging residents in the networking with each other and through the library.

Libraries provide a critical community component where people can resume their normal activities during a disaster. Unlike a large shelter operation that requires visitors to register for service, the library allows for an ebb and flow to maintain control of their life while still receiving assistance. From the stories and observations in Chapter 3, libraries and staff can play a critical role in disaster response and recovery. Their role is often formed through circumstances or serendipity but is emerging to become a standard practice.

Because they are municipal service entities, libraries most often are authorized by state constitutions that designate an advisory or governing board be appointed by municipal officials to provide oversight for library operations. In larger municipalities, libraries fit within or work alongside agencies, such as social services, parks and recreation, or extension services, that provide for recreational, health, or educational services. The municipal government designation can be helpful in forging critical partnerships with like-minded community service agencies. Libraries, like police and fire departments, also stand out as notable community agencies representing public facilities, information technology, cultural programming, and trained and dedicated professionals. The depth and breadth of the library resources are known to the public and that fact makes them unique among municipal government services. As noted previously, they are the go-to place for many reasons including disaster relief.

LIBRARY FACILITIES

As noted previously, public library facilities are used by community residents for a variety of reasons. In the past several years they have become the choice of disaster survivors because of they are familiar and local. Their long standing place in the community is well documented but they are also sturdy, well-built commercial structures intended to accommodate large crowds.

There is a difference between residential and commercial design and construction. According to Forte, architectural design for any facilities is a partnership between the architect and the interior designer. For residential buildings the architect begins the process by ensuring that the home is safe and structurally sound and meets local building codes including regulations for plumbing and electrical safety. Once the basic structure has been designed and is in place the interior designer takes over. They design an interior by selecting textures, colors, and layout that creates a comfortable environment in which to relax and carry out daily tasks.

On paper, it seems that commercial design follows the same process. Architects create buildings with a sound infrastructure that meet local building code regulations; yet there are significant differences between the two. The residential building size and envelope is relatively small when compared to commercial properties. Commercial buildings are usually larger than residential buildings, which demands a greater responsibility for safety and structural integrity. Commercial buildings also have greater infrastructure needs. They will need elevators to allow for freight and people, bathrooms for visitors and employees, cafeterias, and even parking areas. All of these items are less common in a residential building, even if the residential building is a large condominium complex (Forte, 2014). Most public libraries can provide a facility, ample parking lots, essential information services and resources, and trained librarians during an emergency situation. These are all prized, critical assets; but, it is the librarians that are the constant and the most valuable asset.

According to Samantha Montano (2016), since 1950 emergency professionals have gained significant experience and know-how in responding to disasters. Governments, corporations, utility companies, and nonprofit organizations that make up the emergency management system, have become increasingly more professional in their response and reaction to disasters. The lack of coordination of these groups and the capacity of local communities to marshal resources adds constraints to the ability to respond to all needs a community may face during and after a disaster.

The public libraries' response in emergency situations is not new. Public libraries have been responding locally to disaster survivors in a variety of roles and with an array of services and resources oftentimes unbeknownst to local emergency managers and first responders. As noted, survivors find their way to the library during local emergency events. To be more effective in their assumed role, public libraries should forge collaborative partnerships with key local government leaders and departments that are responsible for mounting an emergency response.

Public library leaders are encouraged to collaborate with the emergency manager first. They are typically trained in the art and science of emergency management and apply their learning and training within the local government structure daily. Emergency managers are also responsible for the collaborative relationships needed with other municipal departments and leaders who are traditionally engaged in emergency responses including public safety, human services, and general services agencies.

Emergency management knowledge at the local level has emerged substantially over the past several years. Initially Fire and EMS personnel administered these services. Emergency management professionals with education and training in disaster response are now included in the Fire and EMS structure and generally take the lead role during these disaster situations.

CCPL's experience caused a shift in its official approach during emergency situations from donation and volunteer coordination only to include elements of mass care and public communication. This change required rewriting the local emergency response manual and providing additional training for key library leaders, managers, and line staff. All these actions were easily accomplished due to existing staff knowledge and experience. These internal changes were made with the support and consultation of Emergency Management staff and personnel (Mabe, 2016).

REFERENCES

Becker, C. (2009, March). Disaster recovery: A local government responsibility. *Public Management, 19*(2), 6–12. Retrieved from http://webapps.icma.org/pm/9102/public/cover.cfm?author=christine%20becker&title=disaster%20recovery%3A%20%20a%20local%20government%20responsibility

Blanchard, B. (2007). What is emergency management and what are the principles of emergency management. *EMI Emergency Management Roundtable.* Emmitsburg, MD: EMI.

Federal Emergency Management Agency. (n. d.). *EMUSA local role in emergency management.* Retrieved from training.fema.gov/hiedu/docs/hazdem/session%2018--local%20government%20role.doc

Forte, C. (2014, June 29). The design process: Residential vs. commercial. *Architecture Lab.* Retrieved from http://architecturelab.net/the-design-process-residential-vs-commercial/

Harrington, D. (2001, December). Six trends in library design. *Library Journal*, 2001, 12–14.

International City/County Management Association. (2016, May 12). *Forms of local government structure.* Retrieved from http://icma.org/en/icma/knowledge_network/documents/kn/Document/9135/Forms_of_Local_Government_Structure

International Federation of Library Associations and Institutions. (2001). *The public library service: IFLA/UNESCO guidelines for development.* Munchen: K.G. Saur.

Mabe, M. R. (2016). Libraries to the Rescue. In E. N. Decker & J. A. Townes (Eds.), *Handbook of Research on Disaster Management and Contingency Planning in Modern Libraries* (pp. 494–516). Hershey, PA, USA: IGI Global. doi:10.4018/978-1-4666-8624-3.ch021

Montano, S. (2016, August 23). The Louisiana floods are devastating, and climate change will bring more like them. We're not ready. *Vox News.* Retrieved from http://www.vox.com/2016/8/18/12522036/louisiana-flood-climate-change-emergency-management

National League of Cities. (n. d.). Local government authority. Retrieved from http://www.nlc.org/build-skills-and-networks/resources/cities-101/city-powers/local-government-authority

Rubin, R. E. (2010). *Foundations of library and information science* (3rd ed., pp. 165–170). New York, NY: Neal-Schuman Publishers.

United Nations International Strategy for Disaster Reduction. (2005). Hyogo framework for action 2005-2015: Building the resilience of nations and communities to disasters. *Proceedings of theWorld Conference on Disaster Reduction* (pp. 1-25). Kobe, Hyogo, Japan: International Strategy for Disaster Reduction.

United States Environmental Protection Agency. (2016, August 2). Summary of the Emergency Planning and Community Right-To-Know Act. *Laws and Regulations: US Environmental Protection Agency*. Retrieved from https://www.epa.gov/laws-regulations/summary-emergency-planning-community-right-know-act

Whitehouse. (2016, July 7). *State and Local Government*. Retrieved from http://www.whitehouse.gov/1600/state-and-local-government

Wiegand, W. A. (2011). *Main street public library: Community places and reading spaces in the rural heartland, 1876-1956*. Iowa City, IA: University of Iowa Press.

Chapter 5
Emergency Preparation for the Library and Librarian

The importance of libraries' involvement in disaster response was highlighted in the US in 2010, when the Federal Emergency Management Agency changed its policy to recognize libraries as essential community organizations in areas impacted by a disaster, thus making them eligible for temporary relocation funding in times of disaster or emergency. With this recognition, libraries were acknowledged in the US for their potential to play a major role in disasters. The American Library Association (2013) in its Policy Manual "urges governments at all levels to acknowledge and support the essential role local libraries play in providing e-government and emergency response/recovery services, and to include libraries in relevant legislative or other policy actions" (Hagar, 2014).

For any disaster role played by the library, there is a sense of discomfort felt by all staff members who performs this role. Although the disaster related role in most instances is only slightly different, it is still outside normal circumstances in the day-to-day realm of library service. To ensure they are prepared to perform as expected it is crucial that the library staff be included in all relevant emergency management planning, process, and programs so they can provide input in the development of their own role regarding how the library can best serve their local community during a disaster.

DOI: 10.4018/978-1-5225-2196-9.ch005

Copyright ©2017, IGI Global. Copying or distributing in print or electronic forms without written permission of IGI Global is prohibited.

Through first-hand accounts of librarians living through Hurricane Katrina and other more recent natural disasters, librarians always felt they could do something more but are generally not included in any local formalized planning or training related to the emergency management response. Because of the lack of training and orientation, they were either underutilized or relegated to waiting for the storm to pass (Halsted, Clifton, & Wilson, 2014). If library staff in any of these situations had been CERT trained as a baseline and then been included in local emergency exercise practices, not only would they have been better able to serve storm victims, they would have flourished in their own role.

For most libraries, a general knowledge of these essential skills is needed to feel comfortable in a disaster response. Additionally, knowledge of the characteristics of emergency communication practices, a basic understating of disaster psychology, and the ability to work with emergency management professionals helps to prepare the library to play an improved regular role in an emergency response.

The best way to combat the uncertainty and anxiety of being in a "required role" in an emergency management response is through training and exercise. With proper training, librarians can feel confident in the emergency response role they serve in their community as well as feel empowered to know they are helping residents return to a new version of post catastrophe normal. Fortunately, there are training programs that already exist through emergency management that library staff can take advantage of and Community Emergency Response Team (CERT) training is an excellent place to start.

CERT TRAINING

According to FEMA (2011) CERT was first developed in the early 1990s in California as a strategy to develop a volunteer corps to respond to wildfires. Built on the platform of the incident command structure, the program trained members of the community to assist their own neighbors in basic first aid, search and rescue, disaster psychology, and preparedness planning. The program garnered more attention after the attacks of 9/11 due to President Bush's call to action for volunteers. FEMA's goal is to strengthen community safety and preparedness through increased civic participation.

The seven-week course is intended to provide whole community preparedness with neighbors helping neighbors. Since libraries are so entrenched in the local community, this course is the perfect baseline for educating library staff on emergency reaction. CERT training allows them to serve in their traditional assignment; but be aware of the hazards and vulnerabilities that may affect their areas during disasters and how to respond as a critical support team.

There is no doubt that customers flock to the library during whole community emergency situations. Many news articles have been published noting customers' appreciation for the library and its staff in their times of personal need. Due to the success of CCPL's new found emergency management capacity, library administrators and department managers agreed to enhance the library's expanded role by participating in CERT.

A specialized version of CERT training was offered to CCPL staff by certified county emergency management personnel. The training covered disaster preparedness and emergency response including a mock disaster drill. While all library staff who participated in the training received a CERT certification obligating them to respond when called, an agreement to not mobilize the CERT trained staff away from libraries during disasters was put into place.

The CERT trained librarian has had a positive impact in Chesterfield County. Every CCPL manager and librarian completed the 20 hour courses. Keeping their interest was not a problem. Already highly educated, they are naturals at ingesting new information and keeping the information current. CCPL's experience has been that once the librarians are engaged in this training they want to learn everything possible.

Community based preparedness encourages all community members to be prepared for and respond to the hazards that follow a disaster. Studies have shown that groups working together are more effective in the period following a disaster if they have been trained to respond. Further, if the training is interconnected with the social and political makeup of the community including workplaces, their response as a team is even more effective.

The damage caused by natural disasters can severely disrupt individual and community life. Main concerns are restrictions or disruptions to communications, transportation, and utilities. Experts suggest that residents should be prepared to rely on their own resources for up to three days following an event. Because CERT training covers the basic skills that are important to know in a disaster when emergency services are not available, it makes perfect sense that public library members accept and receive this training.

CERT training is composed of seven units: disaster preparedness, fire safety and utility controls, disaster medical operations, light search and rescue operations, CERT organization, disaster psychology and disaster simulation. Each unit is covered individually over several weeks to ensure an effective training experience.

While all the units are critical to the entire training process, librarians should note the following sections are more likely to be used by them when disaster strikes.

- Disaster Preparedness provides clear understanding of the effects of extreme emergencies on infrastructure and the potential loss of basic services including communication, transportation, utilities and availability of food, water, shelter, fuel, and emergency services.
- Fire Safety and Utility Controls provides a broad based overview of how fires occur, the impact on homes and utilities, how to size up a fire situation, fire extinguishers and fire suppression and hazardous materials.
- Disaster Medical Operation covers life threatening conditions and triage including patient evaluation, and basic treatment.
- Light Search and Rescue addresses how to safely conduct rescue operations.
- CERT Organization shows how to organize and deploy CERT resources, including how to protect your safety and the safety of your CERT assigned buddy.
- Disaster Psychology helps participants to understand the psychological impact of a disaster on you, your peers, the team, and survivors.

Not all disasters result in severe circumstances. Often the basic needs of the survivors are easily met with simple services that libraries provide officially and unofficially. The value of CERT training is that it prepares any group or individual to provide service during and after a disaster.

The benefits of CERT became apparent during several disaster situations in which CCPL staff provided care. In late June 2012, a series of severe storms passed through the Richmond, Virginia area. According to reports filed by Cloak (2012) power was knocked out intermittently for short periods of time in various areas throughout Chesterfield County for several days. The primary problem was not the lack of power but the seasonal high temperatures and an unhealthy ozone level. This combination created dangerous air quality levels, especially for those with acute or chronic respiratory health care concerns. The event required mobilizing some mass care resources although the resources required were minimal.

County EOC staff, consulting with various county leaders, considered several strategies to meet this emergency need without activating mass care level resources. During these strategy sessions, it was noted that library staff updated their response strategies to become part of the mass care ESF including providing temporary daytime shelters for all residents and nighttime shelters for small groups if needed. Additionally, the CERT trained library staff factored into the discussion. CCPL staff and resources were quickly enlisted to render its unique brand of mass care after the freak storm hit the area and impacted the community during the 4th of July holiday.

Because the library was being opened during a major holiday, the EOC arranged for additional CERT volunteers to supplement library staffing at the branches that would be open. The CERT volunteers assigned were pleased to be part of the process and were pleasantly surprised to learn that many of the library staff were also CERT certified.

In addition to the extensive CERT training, first aid and CPR are alternative training options for library staff. Basic first aid and CPR skills are necessary to recognize and treat the symptoms of shock or disaster related stress. These skills are not only important during daily interaction with customers and coworkers that may need assistance during normal hours, but this training also empowers the librarian to assist during a disaster response in a mass care setting. Through past experience, librarians understand that they are often the first contact for some of the disaster victims. They should feel confident and prepared to respond appropriately to health related concerns, especially when directing victims to more specific care.

CERT and first aid training are more successful when there is direct contact between the emergency management program trainers and library staff. A point person can be a helpful, nonattached liaison between the two groups. The liaison should have enough knowledge of emergency management and the library system to be able to facilitate the specific needs and tone of both groups.

Disaster Psychology

There are many kinds of trauma that impact the psychological health of victims and survivors of disasters. Earthquakes carry the highest risk of physical damage where other disasters are less physically stressful. General distress levels following an earthquake return to normal about 12 months later; but, post-traumatic stress reactions do not fade until 18 months after the fact. PTSD levels following earthquakes can reach as high as 95% (Department of Veterans Affairs, 2016).

On the other hand, physical damage due to floods and hurricanes has far less impact. Most survivors of these types of disasters experience only moderate impairment with only 25% of survivors showing symptoms of a diagnosable disorder. Similarly, fires, which have less frequent but more severe psychological impact over the years can cause moderate physical and mental reactions.

In all cases, many common factors determine a given disaster's effects on survivors. Continuing aftershock, loss of homes, income, and even questioning of religious beliefs are some of the factors that can contribute to the mental, emotional, and physical health and well-being of disaster survivors.

A key element of any responder's success is their ability to understand the psychology behind the impact of disasters and disaster scenes and the events related to a survivor's physical and psychological health. Psychology first aid is often needed more than medical first aid.

During a disaster, victims and rescuers are exposed to a variety of situations that are shocking and disturbing. While not necessarily rescuers, librarians providing services during a disaster may be exposed to or provide service for regular customers who are experiencing disaster induced trauma. Recognizing the signs of such trauma and knowing how to react is necessary.

People who survive natural disasters face three phases of traumatic stress reaction. These phases are impact, post-disaster, and recovery. In the impact phase which occurs as the event is initially being experienced, people feel a sense of helplessness and powerlessness due to their inability to stop the impending event and its possible impact. Under the threat of loss of life, being dislocated from families and friends, and even inescapable horror, those who are impacted react in ways to protect their own lives and the lives of others. Feelings of disorganization or being stunned are typical during this phase.

In the post-disaster phase, people tend to recoil from the event as the initial rescue activities commence. People in this phase behave with confusion or demonstrate high anxiety. These feelings are shown in numbness, denial or shock, grief, despair, and sadness. These reactions may vary depending on the individual's perceptions and experience of the elements that cause stress. While some may experience a delay in these reactions, others may show elation due to survival.

A prolonged period of adjustment characterizes the recovery phase. It starts when the rescue is completed and people are starting to bring their lives back to normal. There may be a honeymoon phase due to coming together as a community which can be therapeutic. Disillusionment may follow when the disaster is relived day to day in conversations and news reports.

Because so much attention has been paid to basic needs in the first two phases, and that attention is no longer needed, other needs emerge in the third phases that are existential and psychological. Unfortunately, these types of needs are often ignored, leaving people feeling frustrated and unfulfilled for a prolonged period. It is typical during this phase for people to not express their feelings of distress because so much has already been done on their behalf.

Common behaviors during this phase include sleep disturbance, indigestion, fatigue, relationship concerns or work difficulties. During this final phase, there is generally a period of adjustment that individuals and the community need to experience to completely resolve the overall impact of the event.

TRAINING EXERCISES

Once the training has been acquired, the next step is to continually apply the newly acquired skill sets to stay fresh. While catastrophic disasters splash across the news headlines they are infrequent. Because of this infrequency skills can be forgotten; therefore it is necessary to maintain the partnership with emergency management to be part of the exercise program. Through the Department of Homeland Security's Exercise and Evaluation Program (HSEEP) there is a detailed outline that defines the steps necessary to conduct an exercise as well are create an after action report to evaluate best practices and lessons learned. Through the HSEEP model one will find that there are various levels of exercise training and development. Large full-scale exercises are conducted annually by local emergency management professionals with the FEMA manuals and guidance on the different training practices. Drills, workshops and table top exercises are also a part of the program that need to be practiced more frequently. These exercises can be real world scenarios or simply talking through the unfamiliar concept. Given their skill set and potential use during a disaster, librarians should ensure they are included in all local training exercises.

There are different types of exercises that can be used to evaluate program plans, procedures, and capabilities. The Department of Homeland Security (2016) provides the various elements included in a designated training exercise as well as helpful training programs. The training exercise includes walk-throughs, workshops or orientation seminars, tabletop exercises, functional exercises, and full-scale exercises.

Walkthroughs, workshops, and orientation seminars are basic training for team members. They are designed to familiarize team members with emergency response, business continuity and crisis communications plans, and their roles and responsibilities as defined in the plans.

Tabletop exercises are discussion-based sessions where team members meet in an informal, classroom setting to discuss their roles during an emergency and their responses to a particular emergency situation. A facilitator guides participants through a discussion of one or more scenarios. The duration of a tabletop exercise depends on the audience, the topic, and the exercise objectives. Many tabletop exercises can be conducted in a few hours, so they are cost-effective tools to validate plans and capabilities.

Functional exercises allow personnel to validate plans and readiness by performing their duties in a simulated operational environment. Activities for a functional exercise are scenario-driven, such as the failure of a critical business function or a specific hazard scenario. Functional exercises are designed to exercise specific team members, procedures, and resources (e.g. communications, warning, notifications, and equipment set-up).

A full-scale exercise is as close to the real thing as possible. It is a lengthy exercise which takes place on location using, as much as possible, the equipment and personnel that would be called upon in a real event. Full-scale exercises are conducted by public agencies and they often include participation from local businesses.

Through training and exercise evaluations library and emergency management staff can see how they can best serve the community and the best staffing plan to incorporate.

EMERGENCY COMMUNICATIONS CHARACTERISTICS AND RESOURCES

According to FEMA (2014) during an emergency or disaster incident, there are key communications directed to the community that are critical to ensure a successful emergency response. Emergency communications are unique in that they include:

- Specific alerts and warnings,
- Directives about evacuation, curfews, and other self-protective actions,
- Information about response status, family members, available assistance, and other matters that impact response and recovery.

A well-conceived and effectively delivered emergency message is essential to ensure a successful response. The emergency message may provide public safety information or information to protect property. Depending on the circumstances, the message can be a call for volunteer help with response efforts, a request seeking cooperation or a motivational message to instill public confidence and help families reunite. In all instances, getting the community to respond to the message is the goal.

Experience has shown that many factors influence a community's response to emergency messages. Aside from the message itself, personal perceptions, credibility of the source and the delivery format of the message all influence response. Because of their place in the community, public libraries can be very helpful in adding credibility to an emergency message.

Today there are a wide variety of communication tools to choose from when distributing emergency messages. Each tool has its strengths and weaknesses, depending on the audience you are trying to influence. No one tool is effective for all audiences.

Verbal messages although the most effective are limited by man power and physical distribution factors. Broadcast media is more effective but may be limited due to disrupted reception. Internet and social media also require a receiver to access to message. Regardless of the tool, emergency communications must be clear, contain specific information that does not contradict other or previous information, and be accessible by the entire community.

Emergency communications are automatically disadvantaged due to personal barriers, issues of timeliness, and their need for a response or reaction from their audience. Stress, lack of sleep, and obvious changes in routine make it more difficult for people to hear an emergency message. If the message requires a response or reaction, elements of timeliness become critical in the development and delivery of the message.

FEMA indicates that many emergency message concerns can be alleviated by adhering to the following guidelines:

- Specific – If the message is not specific enough about the "Who? What? When? Where? Why? and How?" the public will spend more time seeking specific information to confirm the risk. If necessary, be specific about what is or is not known about the hazard.
- Consistent – An alert/warning should be internally consistent—that is, one part of the message should not contradict another part. It should be consistent with messages that are distributed via other channels. To the extent possible, alerts/warnings should be consistent from event to event, to the degree that the hazard is similar.
- Certain – Avoid conveying a sense of uncertainty either in content or in tone. Confine the message to what is known, or if necessary, describe what is unknown in certain terms. Do not guess or speculate.
- Clear – Use common words that can be easily understood. Do not use technical terminology or jargon. If protective instructions are precautionary, state so clearly. If the probability of occurrence of the hazard

event is less than 100%, try to convey in simple terms what the likelihood of occurrence is.

- Accurate – Do not overstate or understate the facts. Do not omit important information. Convey respect for the intelligence and judgement of your public (FEMA, 2014).

Errors or omissions in moments of urgency can be prevented with the use of templates tailored to the specific types of threats and hazards unique to your community. Using a template that incorporates pre-approved language can also reduce delays when issuing alerts and warnings. Templates also solve the problems with multi-language communities.

CALLING TREE

To reduce the impact of disasters on communities and to position libraries to be more prepared, the National Library of Medicine through the National Network of Libraries of Medicine (NN/LM) (2016) has developed the Disaster Ready Initiative. This initiative will help all libraries learn how to maintain core information resources and services following a disaster and be prepared to support the health and general well-being of their communities.

Under the leadership of Daniel T. Wilson, Associate Director for Collections & Library Services of the Claude Moore Health Sciences Library at the University of Virginia, the goals and practices of the initiative are being presented at conferences around the county. During their presentations a simple one-page summary plan is offered as a handy tool for library staff to stay informed and connected during a local emergency. The Pocket Response Plan (PReP) was developed by the Council of State Archivists (CoSA) and adapted by the Lodi Public Library in Lodi, New Jersey, and the University of Virginia Claude Moore Health Sciences Library. The goal of the tool is to improve workplace readiness and increase the value of all libraries and librarians in their community during times of disasters.

This tool creates a simple communication matrix that arranges critical information on a single page that can be easily folded into a pocket sized brochure and distributed to all library staff members. The one-page summary includes phone numbers of critical staff and teams, a brief overview of the libraries communications plan, continuity of service plan, operational priorities in case of emergency, floor plans, and other critical information deemed essential to a response by library staff.

Far from a simple calling tree tool, the PReP is a handy, customizable, low tech tool that is easily developed, distributed, and managed at the local level. Public libraries should consider this tool as a starting point in tools and resources for staff and leaders engaged in the emergency response. An example of this tool is included in this volume.

While the calling tree method of dialing a list of numbers is still a viable option, most every locality now has a rapid notification system. These systems allow for databases to be built with pre-scripted messaging that can recall staff or reassure that business is normal. A librarian should inquire within their communications division of government what the notification system is and what staff can be trained to use it. The staff assigned will have the knowledge to quickly get information out or amass resources in preparation for a disaster response.

EMERGENCY ESSENTIALS KIT

Through its Ready Campaign FEMA (n.d.), educates and empowers Americans to take a few simple steps to prepare for and respond to potential emergencies, including natural disasters and terrorist attacks. Ready asks individuals to do three key things: get an emergency supply kit, make a family emergency plan, and be informed about the different types of emergencies that could occur and their appropriate responses. All Americans should have some basic supplies on hand to survive for at least three days if an emergency occurs. They recommend the following basic items be included in a basic emergency supply kit:

- Water and food for at least three days, including a can opener
- Battery powered radio
- Flashlight with extra battcries
- First aid kit
- Local maps
- And other sundry items such as dust mask, moist towelettes, garbage bags, simple tools, etc. (n.d.)

Individuals should also consider where they live and the unique needs of their family or environment and create an emergency supply kit that reflects personal or individual needs. Individuals should also consider having at least two emergency supply kits, one for home and a portable kit for their vehicle or workplace. Public libraries require a similar strategy for staff. A basic kit at work will prepare staff to be on call and ready to serve when the need arises.

This approach can be adapted for the library setting. A basic facility emergency kit should include flashlights, first aid kit, maps, emergency operations manual (or copies of an OPAL plan document). During Hurricane Irene in 2011, library staff noted that the lack of power strips made it impossible to accommodate the large number of customers seeking power to recharge their cell phones and other portable electronic devices. After Irene, each library was outfitted with an emergency essentials kit for staff and customers to accommodate the recharging and other emergency needs. The CCPL basic kit now includes more flash lights, a deluxe first aid kit, multiple power cords, multiple surge protectors, multiple power strips, and dozens of fresh batteries.

It should be noted that people come to the library during emergencies because the library offers a variety of services and resources that supplement the essentials maintained in a personal or family emergency kit. These include water, power, restrooms, access to up-to-date news and information, and internet access. Librarians who are CERT trained or are part of the local emergency response team also provide access to an increased level of medical and information resources.

EMERGENCY MANAGEMENT DOCTRINE AND PHILOSOPHY

Disaster management in the United States has utilized the functional, all-hazard approach for over 20 years now. Rather than developing a single-hazard/threat focused plan such as a tornado plan, emergency managers have embraced this approach that incorporates universal processes used in all disasters such as emergency communications, public warning systems and victim sheltering. Processes, more than single event plans, can be mapped to specific hazards/threats. This allows the manager to look for gaps, overlaps, and conflicts between processes in real time in order to respond more effectively during a disaster and to be better prepared for the next disaster.

With an emphasis on disaster relief and recovery processes rather than single events, emergency management took another positive step forward in the management of natural disasters with the development of the National Response Plan (Department of Homeland Security, n.d.). The NRP is an all-discipline, all-hazards plan that establishes a single, comprehensive framework for the management of domestic incidents. It provides the structure and mechanisms for the coordination of federal support to state, local, and tribal incident managers and for exercising direct federal authority and responsibility. The NRP assists in the important homeland security mission of preventing terrorist attacks within the United States, reducing the vulnerability to all natural and man-made hazards, minimizing the damage and assisting in the recovery from any type of incident that occurs.

An example of a disaster that utilized all federal, state and local services and approaches was Hurricane Katrina in 2006. This disaster resulted in billions of dollars of personal and public damage and caused the loss of over 1,800 lives. According to Wychock (2012) this single storm forced emergency management professionals to realize that such extreme and unpredictable wind and water damage can leave a relentless path of destruction. The sheer size of the storm impacted people, property, and infrastructure in ways that made it difficult to mobilize a response.

WORKING WITH EMERGENCY MANAGEMENT PROFESSIONALS

Emergency managers are now faced with problems they have seldom before confronted. They are expected to understand complex physical and social systems, conduct sophisticated outcomes analyses, and offer long-term solutions to recurring problems. To address these complexities, education in emergency management should include interdisciplinary and holistic degree programs. Programs should employ curriculum that attempt to solve the real-world problems entailed in linking hazards and sustainability. This approach requires new interdisciplinary degree programs that focus on sustainable hazards mitigation (Darlington, n.d.).

All emergency managers know that communities must address the interdependent causes of natural and man-made disasters and come to some decision about which potential risks and losses are acceptable, which are unacceptable, and what specific actions are necessary to maintain the social, economic, and political stability necessary for the community to flourish. They seldom perceive this in the context of a broader role for emergency management in community planning. But consider the connection between the two. For example, if a community is seeking to promote sustainability in the face of serious earthquake risks, structural mitigation alone is insufficient. Much more is required than building codes and the like. Sustainability also requires a linkage of policies on building codes to policies on housing density, to policies on urban transit, to policies on social equality, to policies on environmental quality, to policies on economic development, etc. In other words, the concept of sustainability together links all policies. This linkage includes emergency management policy and brings the emergency management function to the table as a participant in community planning (Schneider, n.d.).

Most emergency management professionals have some degree of public safety experience and training. They also need the capacity to look at the big picture and see simplistic approaches to fixing big problems. During disasters, citizen vulnerabilities to basic services, connectivity, and power are exposed. While there are very few standardized procedures of libraries working in disasters there are many stories that show that it is happening. Emergency management professionals should seek out libraries and partnerships to develop protocols and harness their local libraries capabilities.

PARTNERSHIPS AND COLLABORATION

According to Rose (National Public Radio, 2013), a factor that frequently goes unnoticed during many emergency responses is that public libraries, as trusted providers of information access, particularly access to e-government, are the most logical agency to incorporate in the local emergency management responses. The national government completely failed to help people after Hurricane Katrina, yet, public libraries in and on the edges of the devastation hummed along, providing temporary daytime shelter, information, and basic human aid. The local nature of the response that libraries provided managed to reach members of those communities much better than national or state level responses. Such local response to crises, while vital, are becoming much harder to find outside of public libraries.

On the single principle of local collaboration, so lacking in the Katrina response, as discussed in New Disaster Planning Guide, Carolyn Clancy, MD Director of the Agency for Healthcare Research and Quality, indicated that "to be effective in their planning efforts, local, state and regional leaders needed to be aware of the latest research, tools and models available" (2006). As noted, the lack of collaborative plans that address events at a catastrophic level and the inability of agencies to coordinate and cooperate among themselves led to the problems of immediate relief, including housing, for the victims of Katrina.

Local plans must address recovery actions essential to tackling the short and long term vulnerabilities. Rebuilding the infrastructure, restoring telecommunications, transportation networks, and providing health care and rehabilitation services will all need to be addressed in the aftermath of a disaster. However, pre and post response activities should blend with the development of other initiatives, such as training local response staff including librarians to provide aid in the interim.

WHY LIBRARIES ARE THE PERFECT FIT

According to recent research completed by Pardeep Rattan (2013) on the subject of the importance of the library in disaster management.

The library and information centres world over have experienced a sea change in the methods of acquiring, processing, storing and making that information available to the end users especially after a revolution in the information and communication technologies. Libraries have very successfully geared themselves to adopt any new technology that is emerging with new innovations every day. Libraries are now functional 24x7 day and night in a networked world. In the era of www, internet and other technological advancements that have been implemented in library and information centres, the librarians are now facing a new challenge in managing the affairs of their institutions in the present scenario of new modes and methods of information bases, new formats of storage, and ever changing information seeking behaviour of the users. Librarian is now working as an information scientist, researcher, educator, psychologist, coordinator, public relations officer and what not.

With huge responsibilities on the shoulders of a librarian or the information scientist one has also to safeguard and preserve the variety of precious and invaluable information sources and tools in print and non-print form, staff and other infrastructure from any potential disaster or any risk. These disasters – man made or natural can completely or partially damage the information sources and infrastructure kept and preserved for the use of present generation and for posterity. At the same time it can also paralyse the functioning of that particular library and information centre where the disaster has hit. It is in this light why planning to face any potential disaster is so important (2013).

Rattan's research makes the obvious point that today's librarian is much more than a book clerk and the library in the community is more than a warehouse for books.

In most of the cases the librarians and the supporting staff is ready to assume any role and help the community during any disaster. 100% of the respondents are ready to act as a support for their institution, to act as coworker, to share and disseminate the relevant information regarding any disaster through web.

A majority of them around 86% agree to work with local administration to help in rehabilitation and recovery from the emergency. Although all the libraries are equipped with firefighting instruments but a majority of the staff is not sure of its handling, even they have not checked its expiry dates. The lack of any special budgetary provisions is a cause of concern because without finances at the discretion of the library administrator, an immediate action to tackle the emergency cannot be taken up. Another point is of provision of any training or instructions for the working professionals to protect self, their resources, community, and to educate common masses regarding potential risks from any disaster and the measures to safeguard themselves against the disasters. One most glaring outcome of the study shows that only 29% of the respondents agreed to have a disaster plan in their libraries and a large majority 71% even are not aware if there is any such plan which addresses the issue of disaster management (2013).

According to Newman and Newman (2015), while a library should have a plan specific to system disasters, no organization including a library can operate as an island when it comes to natural and man-made disasters that impact the entire community. Likewise, no community can rely on outside help to provide all needed local help. Understanding your place within the broader community and being prepared to act when mobilized is more effective.

In Emily Ashley's experience as the emergency management coordinator for Chesterfield County, there is no official protocol for how a library should respond during a disaster or what their absolute role may be. Through news reports and observations, libraries demonstrate that they are capable of responding. During an interview in June, 2007 with Kay Due, Manager of Public Services at the Memphis Public Library, she describes the various emergency response duties she and her staff performed. Her comments demonstrate the need for librarians to be aware of first aid skills, understanding and distributing emergency communications, understanding the phases of disaster psychology and be familiar with emergency management staff and protocols.

Aside from being available, information savvy, customer service oriented, educated and knowledgeable librarians are well suited to be part of the unknown factors that tend to find their way into emergency response. These unknown factors stem from the fact that communities are still learning how to respond to disaster situations and have not yet harnessed their ability to manage serendipity. But we can understand and plan for it.

REFERENCES

Agency for Healthcare Quality and Research. (2006). *New planning guide offers valuable mass casualty care information.* Retrieved from https://archive.ahrq.gov/news/press/pr2006/mcepr.htm

Cloak, M. (2012, July 4). Life in the summer without power is no fun. *Richmond Times Dispatch*, p. 1A.

Darlington, J. D. (n. d.). *The profession of emergency management: Educational opportunities and gaps.* Washington, DC: Federal Emergency Management Agency. Retrieved from https://training.fema.gov/hiedu/downloads/prof_em.pdf

Department of Homeland Security. (2016). *Exercises: Ready.gov*. Washington, DC: Department of Homeland Security. Retrieved from https://www.ready.gov/business/testing/exercises

Department of Homeland Security. (n. d.). *National response plan: Department of Homeland Security.* Washington, DC: Department of Homeland Security. Retrieved from https://www.dhs.gov/xlibrary/assets/NRP_Brochure.pdf

Department of Veterans Affairs. (2016, May 16). *Traumatic effects of specific types of disasters.* Washington, DC: Department of Veterans Affairs. Retrieved from http://www.ptsd.va.gov/professional/trauma/disaster-terrorism/traumatic-effects-disasters.asp

Federal Emergency Management Agency. (2011, January). *CERT training materials*. Washington, DC: Federal Emergency Management Agency. Retrieved from https://www.fema.gov/training-materials

Federal Emergency Management Agency. (2013, April). Stafford Act: Robert T. Stafford Disaster Relief and Emergency Assistance Act, as amended April 2013. Washington, DC: Federal Emergency Management Agency. Retrieved from https://www.fema.gov/media-library-data/1383153669955-21f970b19e8eaa67087b7da9f4af706e/stafford_act_booklet_042213_508e.pdf

Federal Emergency Management Agency. (2014, February). *Communicating in an emergency.* Washington, DC: Federal Emergency Management Agency. Retrieved from https://training.fema.gov/emiweb/is/is242b/student%20manual/sm_03.pdf

Federal Emergency Management Agency. (n. d.). *Emergency Supply List.* Retrieved from https://www.fema.gov/media-library-data/1390846764394-dc08e309debe561d866b05ac84daf1ee/checklist_2014.pdf

Hagar, C. (2014). The US public library response to natural disasters: A whole community approach. *World Libraries*. Retrieved from http://worldlibraries.dom.edu/index.php/worldlib/article/view/548/472

Halsted, D. D., Clifton, S., & Wilson, D. (2014). *Library as a safe haven: Disaster planning, response, and recovery; A how=to-do-it manual for librarians*. Chicago, IL: ALA Neal-Schuman.

National Network of Libraries of Medicine. (2016). *Emergency preparedness: National Network of Libraries of Medicine*. Retrieved from https://nnlm.gov/ep/category/nnlm-national-emergency-preparedness-plan/

National Public Radio. (2013, August 12). For disaster preparedness: Pack a library card? *Morning Edition, National Public Radio*. Retrieved from http://www.npr.org/2013/08/12/210541233/for-disasters-pack-a-first-aid-kit-bottled-water-and-a-library-card

Newman, D., & Newman, N. (2015). Are disaster and emergency plans truly complete? *Library Leadership & Management*, *39*(3), 1–11.

Rattan, P. (2013). Role of library and information centres in disaster management. *Library Philosophy and Practice (e-journal)*. Retrieved from http://digitalcommons.unl.edu/cgi/viewcontent.cgi?article=2204&context=libph ilprac

Schneider, R. O. (n.d.). *A strategic overview of the "new" emergency management.* Retrieved from https://training.fema.gov/hiedu/downloads/schneiderstrategicoverview.doc

Witteveen, A. (2015, December 29). Public libraries support refugees. *Library Journal.* Retrieved from http://lj.libraryjournal.com/2015/12/public-services/public-libraries-support-refugees/#_

Wychock, E. (2012). Hurricane Katrina: Overview and lessons learned. *Wikispaces*. Retrieved from https://failures.wikispaces.com/Hurricane+Katrina+-+Overview+%26+Lessons+Learned

Chapter 6
Mitigation Planning and Hazard Analysis

Although this volume is directed toward public library professionals and leaders to help them become better prepared for the developing role they play in local emergency management response, this chapter is intended to provide a broad understanding of hazard analysis and mitigation planning; two key areas unique to the profession of emergency management.

Each of these areas is predominately prescribed and monitored by FEMA. Assuming that the local response is the best response, FEMA's oversight and leadership in these areas greatly increases the likelihood of local responses being positive and reduces the likelihood that federal resources are spent disproportionately at the local level.

HAZARD ANALYSIS THIRA MODELS

According to the Department of Homeland Security (2013a), every community should understand the risks it faces. By understanding its risks, a community can make smart decisions about how to manage risk, including developing needed capabilities. Risk is the potential for an unwanted outcome resulting from an incident, event, or occurrence, as determined by its likelihood and the associated consequences. By considering changes to these elements, a community can determine the best way to best manage and plan for the greatest risks across the full range of the threats and hazards it faces.

DOI: 10.4018/978-1-5225-2196-9.ch006

Copyright ©2017, IGI Global. Copying or distributing in print or electronic forms without written permission of IGI Global is prohibited.

The Disaster Mitigation Act of 2000 was the first step in requiring localities to know the hazards they face. The act mandated that each locality must complete a multi-step mitigation plan in order to be eligible to receive disaster relief as outlined by the Robert T. Stafford Disaster Relief and Emergency Assistance Act (FEMA, 2013b). The most critical element to the hazard mitigation plan was a comprehensive Hazard Identification Risk Assessment (HIRA). This risk assessment is comprised of four basic components that identify the hazard, profile historic events, outline assets a community has to combat the hazard, and provide for an estimate of losses (FEMA, 2013b). These four considerations contribute to a vulnerability assessment that can then identify gaps in mitigation and planning efforts. The HIRA is usually based on natural disasters covering only drought, earthquake, flood, hurricane, and wildfires. Considering the man-made hazards and terrorist threats of modern times, the HIRA may seem incomplete. Recognizing this, in 2012 FEMA released the first version of the Comprehensive Preparedness Guide (CPG). A separate entity from a Hazard Mitigation Plan, the CPG presented basic steps in completing a Threat and Hazard Identification and Risk Assessment (THIRA)

While mitigation plans mostly explore natural hazards, a THIRA goes into more detail to examine the following three categories:

- Natural hazards resulting from acts of nature
- Technical hazards resulting from accidents or failure of systems and structures
- Human-caused incidents resulting from intentional adversarial acts

The THIRA process helps communities identify the capability targets and resource requirements needed to address anticipated and unanticipated risks (Department of Homeland Security, 2013).

The THIRA Process is a four-step process that identifies the threat or hazard concern, places the threat or hazard in context, assesses the capability of the hazard or concern, and finally defines resources required to respond to the hazard or event. This process standardizes the risk analysis that emergency managers use every day in determining the community's preparedness to respond to disasters in the five mission areas: prevention, protection, mitigation, response, and recovery.

The Department of Homeland Security (2013) suggests active involvement from the entire community in developing an effective THIRA. When the community realizes that preparedness and reaction is a shared responsibility involving everyone, and not just the domain of emergency management public safety professionals, local responses can be prepared and executed more effectively. With the understanding that every disaster starts locally, and every disaster can start in a home or neighborhood, whole community preparedness is paramount. Compared to a HIRA that gives perspective on disasters, a THIRA engages an entire community for a proactive conversation about what to prepare for and what sharable resources are required.

Local communities should include all representative parties in developing and updating their THIRAs, including the local public library as a critical resource known to the community. An informed public is the best advocate for building and sustaining required capabilities and creating a secure and resilient community.

Two factors are used to determine the need to prepare for a specific hazard or threat:

- The likelihood of an incident or event based on history and available intelligence. Communities should not rely exclusively on historical facts. The local emergency planning committee is an excellent organization to identify the likelihood of an event because public safety officials from several branches of local, state, and federal government that are part of these committees may provide insight on events that occur less often than historical records show. Emergency managers are also knowledgeable about weather patterns and can provide insight on vulnerabilities.
- The significance of the event on the local community. Incidents vary in impact.

Communities should plan and prepare for those incidents that are locally prevalent and significant. Local emergency management professionals will guide the involved community representatives through the execution process. Most disasters can be grouped by areas that have the same vulnerabilities. So while their likelihood may be slim, the public can be prepared for other disasters with a higher threat ranking. For example, the public is trained to shelter indoors away from windows during tornados. Consider a biological attack which has a slim threat assessment; the action is the same. Addressing vulnerabilities rather than hazards is what makes the THIRA such a powerful tool.

MITIGATION PLANNING: REDUCING VULNERABILITIES

According to FEMA's Local Mitigation Planning Handbook (2013a), "Disasters can cause loss of life; damage buildings and infrastructure; and have devastating consequences for a community's economic, social, and environmental well-being." Hazard mitigation reduces disaster damages and is defined as sustained action taken to reduce or eliminate the long-term risk to human life and property from hazards. Outreach programs increase risk awareness. Mitigation plans have a chapter that is dedicated to the goals and implementation of mitigation programs in a community. At the government level, projects to protect critical facilities, and the removal of structures from flood hazard areas are examples of mitigation actions. For a mitigation plan to be successful, it's concepts should be incorporated into land use plans and building codes.

FEMA (2013b) also states that The Robert T. Stafford Disaster Relief and Emergency Assistance Act (Stafford Act), as amended by the Disaster Mitigation Act of 2000, provides the legal basis for state, tribal, and local governments to undertake risk-based approaches to reducing natural hazard risks through mitigation planning. Specifically, the Stafford Act requires state, tribal, and local governments to develop and adopt FEMA-approved hazard mitigation plans as a condition for receiving certain types of non-emergency disaster assistance.

Local governments have the responsibility to protect the health, safety, and welfare of their residents. Proactive mitigation policies and actions help reduce risk and create safer, more disaster resilient communities. Mitigation is an investment in your community's future safety and sustainability. Consider the critical importance of mitigation to:

- Protect public safety and prevent injury and loss of life.
- Reduce harm to existing and future development.
- Prevent damage to a community's unique economic, cultural, and environmental assets.
- Minimize operational downtime and accelerate recovery of government and business after disasters.
- Reduce the costs of disaster response and recovery, and the exposure to risk for first responders.
- Help accomplish other community objectives, such as leveraging capital improvements, infrastructure protection, open space preservation, and economic resiliency (2013a).

FEMA's (2013a) definition of mitigation planning states that the purpose of hazard mitigation is to reduce the potential losses from future disasters. The intent of mitigation planning is to maintain a process that leads to hazard mitigation actions. Mitigation plans identify the hazards that impact communities and identify actions to reduce losses from those hazards.

There are five guiding principles promoted by FEMA to assist communities in the development of a local mitigation plan. FEMA uses these principles when evaluating local plans for compliance. FEMA promotes a performance based plan rather than a prescriptive approach. Their advice is more along the lines of what should be done rather than how it should be done. When developing mitigation plan, communities should focus on mitigation strategy, compliance, process, community specifics, and relationships. Each principle is briefly reviewed below.

The mitigation strategy provides a long view for reducing loss locally. This principle calls for goals and long term policy statements from a 'global' view and ensures that local policies, programs, and resources are available for use or modification in order to reduce vulnerability and loss.

In order to be eligible for federal disaster relief funds, the local mitigation plan must comply with FEMA rules and guidelines in the area of preplanning, risk assessment, and mitigation strategy. Most often these mitigation plans are completed within a regional scope with the understanding that larger disasters cover more area and may cross jurisdictional boundaries.

A plan that is compliant meets all necessary standards to demonstrate the community's overall intent to provide effective disaster mitigation. Libraries can be seen as part of that community's overall approach to provide effective disaster mitigation by hosting outreach preparation programs and participating in the post-disaster recovery process.

All communities are unique in the type of disaster they may face and the level of response they may mount. However, FEMA looks for evidence that the mitigation plan represents the community and is committed to disaster mitigation regardless of the size or the local resources allotted towards mitigation. In many of the examples put forward in this volume, local public libraries and library staff were key resources that provided an unanticipated level of mitigation to survivors.

A mitigation plan is necessary for access to FEMA resources and the plan must comply with FEMA principles. In the area of relationships, FEMA expects local mitigation plans to show clear working relationships with state and community partners and that the partners included in the plan are aware of the plans implications for them. As indicated previously, public library responses are varied. Not every facility or staff member can render assistance;

however, being included in the mitigation planning process helps define the type of partner they will be in future disasters.

While developing a mitigation plan, FEMA does offer assistance in terms of classes for emergency managers to learn, such as the Hazard Mitigation Plan course through the National Emergency Management Institute and also grant funding to provide hired expertise to locals in their plan development. Through these methods, one will learn that there is a logical process in plan development. The planning process naturally includes all officials, partners, and stakeholders in the mitigation process, ensuring that the plan is community-driven and shows clear evidence of broad discussion and documentation of recommended actions.

The primary purpose of a mitigation plan is to ensure communities can manage mitigation locally as much as possible before state and federal resources are provided. The overall success of a plan is relative. While the goal of mitigation planning is to reduce loss of life and limb, and to minimize structural and infrastructural damage, no plan can change the nature or course of a disaster. So how can you gauge the success of a mitigation plan?

Local government comprehensive plans and claims paid by property insurance companies for losses due to weather-related natural disasters between 1994 and 2000 totaled more than $26 billion (in constant 2000 dollars) for disaster-related claims arising from losses to residential property. Although a majority of states do not require local governments to prepare comprehensive plans, 24 states do require plans, and 10 states specifically require that mandated plans pay attention to natural hazards. Multivariate analyses indicate that insured losses to residential property over the period studied could have been reduced by 0.52% if all states had required local comprehensive plans and by a further 0.47% if, in addition, they had required consideration of natural hazards in local plans. Over the period studied, if all states had required comprehensive plans with hazard mitigation elements, the total in insured losses to residential property from natural disasters would have been reduced by approximately $213 million in constant 2000 dollars (Burby, 2005).

Mitigation plans can be viewed as successful by evaluating the Strategic Measures within the plan. These outlined goals and objectives are attainable within the four-year life span of a plan. If a community is truly engaged in the mitigation process the plan will be traced to other plans or public safety minded measures. The stated goals in the section are easy to identify and show an impactful after action post implementation.

MODES OF ADAPTATION: WHAT CAN WE EXPECT?

Research suggests that there are modes of adaptation inter-organizational systems use when responding to extreme events such as earthquakes or other natural disasters (Comfort, 2002). This model fits nicely with public libraries that have emergency experience and can be used during mock disasters or preplanning events before they accept an invitation to be part of a municipal emergency response plan.

There are four modes of adaptation based on the conditions of the organization's technical, organizational, and cultural systems before, during and after an extreme event. The technical factors measure availability of infrastructure such as transportation, electrical power, and communications. Organizational factors measure the flexibility and adaptability of the organization, communication styles, and leadership. Cultural factors measure openness to innovation, willingness to accept new concepts or initiate new patterns of action (Comfort, 2002). The four modes of adaptation are:

- Non-adaptive systems demonstrate low technical, organizational, and cultural factors and function largely with outside assistance.
- Emergent systems demonstrate low technical factors and medium flexibility and cultural openness. They function and learn to cope, but are not able to sustain their approach.
- Operative systems demonstrate medium technical, organizational, and cultural factors and operate well, but are unable to translate their approach into new modes of sustained reduction.
- Auto-adaptive systems measure high on technical, operational, and cultural factors and respond well and can transfer lessons learned into sustained reduction.

All types of systems can provide assistance. However, auto-adaptive systems are best suited for extreme events. An auto-adaptive system tends to move through five phases: information searching; information exchange; sense making; actual adaptation based on choices made; and evaluation (Comfort, 2002). These phases mirror the librarian's approach with a customer seeking data and information.

Since emergency management professionals do not typically maintain the staff resources they control in mounting a response to an emergency, coordination among the various departments with unique skill sets is critical. These two approaches to mitigation analysis can assist library staff and administrators, as well as other nonemergency management departments, by providing them with a common language and established approach for developing their roles in response to local emergency situations. For libraries with designated emergency roles, the THIRA model provides a clear picture of each emergency and the likelihood of library mobilization. Likewise, the auto-adaptive model can aid library leaders and staff with a practical way to evaluate the stages of an emergency and know what to expect. While these models are helpful in preparing for and participating in an emergency response, they do not take the place of a local emergency response plan.

REFERENCES

Burby, R. (2005). Have state comprehensive planning mandates reduced insured losses from natural disaster? *Natural Hazards Review, 6*(2), 67-81. doi:10.1061/(ASCE)1527-6988(2005)6:2(67)

Comfort, L. (2002). Managing intergovernmental responses to terrorism and other extreme events. *Publius, 32*(4), 29–29. doi:10.1093/oxfordjournals.pubjof.a004970

Department of Homeland Security. (2013, August). *Threat and hazard identification and risk assessment guide: Comprehensive preparedness guide (CPG) 201 2nd edition*. Washington, DC: Department of Homeland Security. Retrieved from https://www.fema.gov/threat-and-hazard-identification-and-risk-assessment

Federal Emergency Management Agency. (2013a, March). *Local mitigation planning handbook*. Washington, DC: Federal Emergency Management Agency. Retrieved from https://www.fema.gov/media-library-data/20130726-1910-25045-9160/fema_local_mitigation_handbook.pdf

Federal Emergency Management Agency. (2013b, April). *Stafford Act: Robert T. Stafford Disaster Relief and Emergency Assistance Act, as amended April 2013*. Washington, DC: Federal Emergency Management Agency. Retrieved from https://www.fema.gov/media-library-data/1383153669955-21f970b19e8eaa67087b7da9f4af706e/stafford_act_booklet_042213_508e

Chapter 7
I'm Trained, Now What?

Throughout this text there has been ample information and examples to show that with an engaged partnership, libraries and emergency management departments have the tools and resources needed for a highly formative relationship that can provide rarely used but valuable services to survivors and first responders during disaster response. In the article, "Connecting Public Libraries with Community Emergency Responders" (Malizia, Hamilton, Littrell, Vargas, & Olney, 2012), the authors discuss the 2009 National Network of Libraries, South Central Region (NN/LM SCR) network offices' project that explored ways to support emergency responses of public libraries to natural disasters. Eighteen libraries in Louisiana and Texas, states hard hit by natural disaster, were included in interviews and focus groups to define lessons learned that could be generalized to public libraries that face a variety of disasters. The following insights were gained in promoting the public library as a key resource to emergency responders:

- Library staff are skilled at assisting the general public in locating information.
- Librarians are masters at mobilizing and distributing information.
- Public libraries are comforting places for disaster victims.
- Multi-purpose spaces in public libraries can be adapted for many uses.
- Most public libraries are wired for internet connectivity.
- Libraries serve more than local residents.

DOI: 10.4018/978-1-5225-2196-9.ch007

Copyright ©2017, IGI Global. Copying or distributing in print or electronic forms without written permission of IGI Global is prohibited.

In addition to these factors that are directly related to librarians and library facilities, other insights were noted. With emergency professionals, libraries are available to host and support informational/educational events for the public on a number of topics including:

- Health information resources
- Mental health information
- Printed information during emergencies
- Networking assistance with emergency professionals

According to Bishop, other surveys have identified the following insights about the natural role of the library as an information hub (Bishop & Veil, 2013):

- A source of power for charging cell phones and laptops
- Places of collaboration for FEMA and Red Cross service
- Location for portable toilets
- Drop off and pick up site for donations and resources

In all instances, when the public library assumes a visible role, library leaders must meet regularly with emergency managers to ensure efficient cooperation. There are many examples of libraries playing a natural role during a disaster event because they have a facility and essential services. The partnerships and staff value is substantially enhanced as library staff acquire and incorporate essential disaster response training. After training, both entities must take action.

LOGICAL RESPONSE SCENARIOS

It is critical that a library-emergency management partnership be outlined in the local Emergency Operations Plan (EOP). Standard operating procedures in the EOP and the library's policies and procedures should be developed to be robust when defining viable response scenario options for engagement during disasters, including appropriate application and maintenance of requisite training between emergencies. Standard operating procedures should be supported by an overall assessment that defines the best use of facility space, library staff strengths, library facility location within the community, parking lot size and volume. For library systems with multiple branches or cooperative regions served by several libraries, this assessment should include all staff and facility locations in the system or region. Based on this assessment library and emergency professionals can then predetermine the best use of its library staff, facility, and its resources.

The library's usage is best described within the four anticipated response scenarios exhibited during recent disaster and emergency situations. These scenarios include Pre-disaster Education and Outreach Center, Comfort Stations, Family Assistance Centers, and Volunteer and Donations Center Management. The examples of library disaster response throughout this volume fall mainly within the Comfort Station scenario. Most libraries cited provided warming/cooling stations, restrooms, and charging stations immediately after emergency and disaster situations. A few libraries have provided Family Assistance Center services and even fewer libraries have provided Volunteer and Donation Management. For the scenarios to work and be successful, plans and training must be put in place before any disaster happens. Libraries that previously provided service possess a laboratory experience to draw upon when making the decision to officially engage in emergency management responses. The four main response scenarios to be used at a library facility and with trained staff are described below. The selection of a scenario should be done after careful assessment and planning.

The first scenario, Pre-Disaster Public Outreach and Education Center, occurs during non-disaster times. This scenario allows the public library to assure the community that they can rely on the library to be a place for disaster information and staff knowledge to aid and assist them when things go wrong.

Using the library as a place of public outreach and education allows the library to demonstrate its importance during a disaster response. Outreach programs should include local emergency management staff and coincide with the height of each disaster season - summer for hurricanes and winter for snow storms, etc. Since each community is different, emergency management staff presentations should be tailored to fit the needs of the community. The unique urban, suburban, and rural nature of the local community should be represented in the information presented. Language variations must be considered in the verbal presentations and written documentation.

Since the community naturally gravitates to the public library during emergency situations, having the outreach events at the library builds trust and increases confidence between customers and staff. Customers understand what services they can expect at the library during a disaster and give valuable feedback about their needs.

The second response scenario, Comfort Station, is the one most commonly used by public libraries. This scenario is easily selected and planned for based on threat analysis. The largest threat for nearly all communities is a loss of power. The reasons for these outages vary; but, the threat always remains the same. When a community suffers a large power outage residents require shelter from extreme weather and access to electricity. This access allows them to charge personal devices, use the internet to pay bills or conduct business, connect with concerned family and friends, and get the latest information and possible disaster assistance. While not an actual "shelter," the library can become a unique Comfort Station that will answer the call for basic yet essential services.

Chesterfield County unexpectedly learned about this need during Hurricane Irene when daily library use increased substantially during the time when large portions of the population were without power. Realizing that residents were flocking to the library to get information, plug in their devices, and cool off, the Emergency Operations Center had an ah-ha moment that the library was the best place for mass feeding.

While past experiences have had the Operations Center scrambling for a space from the faith-based community for these services, the library became a serendipitous solution for many reasons. Most importantly, county emergency management staff and resources would not have to be tasked out of the area to monitor security or provide manpower for these required service events. Event security could be provided by local law enforcement. Emergency management leaders realized that the spacious library parking areas were able to accommodate mass feeding.

With these considerations, CCPL worked flawlessly as a mass care feeding site during the Hurricane Irene event. Emergency management staff soon realized that because library staff was already skilled at material handling processes, it was not a stretch for them to process prepackaged meals. Librarians are well schooled in the art and science of answering reference queries, so answering the frequently asked disaster questions would be business as usual. This phenomenon was repeated in the summer of 2012 when the EOC observed that people voted with their feet for their local library to be their "go to" place for service and information.

In late June 2012, a derecho, which is a strong straight-line wind storm, impacted Chesterfield County. Much of the county was without power for several days and with the Fourth of July holiday weekend looming it was not prudent for the county to offer the typical limited holiday services. By this time, library staff had been trained in the Citizen Emergency Response Team (CERT) program and a library emergency staffing plan was in place. The training in conjunction with the staffing plan enabled county emergency management staff and leaders to open libraries in highly impacted areas during scheduled closings, extend the hours, and supplement the library staff with community volunteers. This event standardized the practice of CCPL libraries being a dependable and reliable Comfort Station for people and public safety staff in Chesterfield County.

The third scenario, Family Assistance Center, arises out of the threat environment of 2016. Community organizations should consider active threats such as mass shootings to be real and emergent disasters. The same trends materialized in every published after action report of these horrific events. The trend demonstrates a failure in communications due to systems being overwhelmed and the community not having a pre-identified location for a Family Assistance Center. While some communities offered a reunification center as an area to quickly return people involved in an active threat scenario to their loved ones, a Family Assistance Center is a much more personal and intimate setting for worst case scenario information to be processed.

Public libraries are the ideal location to host these centers. With staff trained in CERT and disaster psychology, librarians offer the empathy to serve as logistics managers for other resource personnel to come in and manage the actual victims. The infrastructure of the library is ideal because the layout frequently offers rooms of various sizes that can used for breakout sessions, mortuary services, advocacy or legal information, and one-on-one counseling. Larger meeting rooms can serve as briefing areas for family members. Library offices are ideal for staff to file case information and provide assigned staff access to phone lines or technology needs. Most importantly, a library can be dedicated as a Family Assistance Center for any amount of time. For as long as necessary, residents can receive normal services from other library locations in the region so that victims and the community can access the library Family Assistance Center to mourn and start the healing process.

Experience suggests that it is much easier to secure a location that is owned by the municipality rather than relying on volunteered space or a private or non-profit partner. To be successful in a Family Assistance Center, pre-planning is crucial. A team of emergency management professionals, librarians, public safety, and mental health staff must do a site visit during non-disaster times. These site visits allow the team to talk through multiple scenarios, mark a proposed layout for the facility for best uses, and identify gaps to be addressed.

In a Family Assistance Center plan, librarians never take the primary lead; however, they play critical roles in the facilitation of various groups of professionals that serve victims. Libraries will be the coordinating unit of the facility to make sure logistical and technology demands are being met and to bring calm to the chaos. The success of a plan is dependent on pre-training which includes disaster drills and table top exercises. The emotional strain of a Family Assistance Center requires evaluating the need for staff critical incident stress debriefings and rotating staff through shorter shifts.

The experience of Ferguson Municipal Public Library during their month of turmoil is a serendipitous example of library professionals taking action. Sensing the fear and concern of the community, they provided a Family Assistance Center-like service. The director's decision to keep the library open as a place of refuge from the violence was bold. In this instance the library conveyed its importance as a safe place for all county residents but especially for those in fear.

Similarly, the Orange County Public Library in Silverado, CA operated as a Family Assistance Center-like location during the wildfires of 2008. By providing access to vital information, librarians alleviated the displaced population's frustration in their time of need.

Libraries with experiences like Ferguson Municipal and Orange County Public Libraries are well positioned to work with local emergency management and human resources partners to complete the requisite assessment and become the designated local Family Assistance Center during times of threat induced crisis.

Often when a localized disaster occurs in a community, those not affected want to give back. This can quickly lead to spontaneous volunteers and unsolicited donations becoming a bigger disaster than the actual man-made or weather impact. This little-known aspect of disasters is an opportunity for the library to engage in or select from the fourth response scenario, Volunteer and Donations Management Center. The library is naturally suited for this response scenario.

During Hurricane Sandy, Dan Porter (2012) reporting for Business Insider identified many ways that technology could help manage volunteers and donations. Feeling a need to donate resources, he and his wife headed out early one weekend to find a place to donate their wares. Porter took note of several real-time examples of inefficiency in managing donations and volunteers, but noticed one positive example. At the local community center in Far Rockaway, he found long lines of people waiting to sift through disorganized mounds of donated goods. Already feeling discouraged, Porter and his wife took the advice of a community center volunteer and investigated options at the public library a block away. Once inside the library they found donations well organized and the distribution operation running efficiently. "In only the way brilliant librarians could, they most likely served thousands of people quickly and efficiently while other locales were serving only hundreds with long lines. They did what they knew how to do and were trained to do – but with relief supplies instead of books" (Porter, 2012).

From volunteers and donations to outreach and education, each of these response scenarios can work for an organization. And while these examples occurred in a serendipitous environment, with pre-planning, analysis and established partnerships, the unforeseen moment can shift from meeting needs as they arise to a well-planned process that will find greater success in the face of disaster. The Greek poet Archilochos wrote "We don't rise to the level of our expectations, we fall to the level of our training." When considering best use of facilities and trained staff, both expectation and training should be high.

EMERGENCY OPERATIONS CENTER PARTICIPATION

Of all elements to be considered when developing a role in emergencies, the key goal is to be included in the local EOC. CCPL was invited to participate in the EOC operation based on the skills and talents represented by the profession. Librarians, in addition to being problem solvers and information seekers, are also exceptional communicators. The primary responsibility of CCPL initially was to answer phones and respond to questions. CCPL's experience of becoming part of the process would not have happened had library staff not been in the room to initiate this collaboration.

The EOC is a fast-paced operation and increases in intensity depending on the severity of the event. Librarians in most instances are well suited to answer the phones and solve problems, a key service offered by the EOC. But, depending on the community, libraries can provide help and resources naturally because they are known to their community. It is imperative that key decision makers engage in discussions about the service and resources libraries already offered during disaster events, and what if any additional service they can offer. Applying the auto-adaptive model to initiate these discussions before an event would be best, but learning during an event by being part of the EOC, as was CCPL's experience, is still a viable option.

In 2011 during Hurricane Irene, Chesterfield County staff and departments including CCPL that made up the emergency response team, shifted in and out of these various modes of adaptation, finally arriving at the auto-adaptive mode. Post analysis demonstrates Chesterfield shifted through the five trends of the auto-adaptive mode starting with the data collection process during the daily EOC briefings.

Of special note is CCPL's daily customer count offerings that came through the nine branches in the system after the storm subsided. During team discussions, the realization that more than 1,000 customers per day were visiting each of the nine branch libraries painted a clear picture of the trust and confidence customers placed in the library. As the discussion ensued, the logic of using the library as a mass care feeding distribution point began to set in. A further realization that this was not an unusual volume of customers for library staff, and that they were already capable of managing crowds with needs, further confirmed for other mass care professionals that the recommendation to use the library as a feeding station was not out of the question. Naturally following the five steps in the modes of adaptation model, CCPL was tasked with distributing meals from the branch library most centrally located to the customers in need for four days following Hurricane Irene. The library administration participation in the EOC allowed emergency management to make a decision that best served those impacted by Hurricane Irene.

FINAL WORD

CCPL and other libraries experienced serendipitous moments in conjunction with responding to disaster needs. A few public libraries acted officially, but most acted unofficially. CCPL has established a model based on trial and error experience in their laboratory of local disaster response. During the various disaster events they presented a willing attitude that garnered increased trust in their partnership with Chesterfield County Emergency Management Professionals and county leaders. The result was an emerging trend that put in motion a process that would ensure libraries had a defined role in serving human services needs before, during and after a disaster. And with that role came training, support and inclusion that put the serendipity under their control.

This shift from unofficial to official inclusion required rewriting county emergency response manuals and establishing a collaborative library emergency response manual. The library manual provides branch-specific response details, approaches, and designated training expectations for key library leaders and managers as well as most public-facing line staff. These internal changes were made with the support and consultation of Emergency Management staff who arranged and presented a library specific version of the CERT training and reviewed all internal library documents dealing with library processes and procedures related to emergency response.

REFERENCES

Bishop, B. W., & Veil, S. R. (2013). Public libraries as post-crisis information hubs. *Public Library Quarterly*, *32*(1), 33–44. doi:10.1080/01616846.2013.760390

Malizia, M., & Hamilton, R. Littrell, D., Vargas, k. & Olney, C. (2012, May/June). Connecting public libraries with community emergency responders. *Public Libraries Online.* Retrieved from http://publiclibrariesonline.org/2013/04/emergency_responders/

Porter, D. (2012, November). Hurricane Sandy revealed some huge problems tech founders need to start solving. *Business Insider*. Retrieved from http://www.businessinsider.com/hurricane-sandy-reveals-new-startup-opportunities-2012-11

Related Readings

To continue IGI Global's long-standing tradition of advancing innovation through emerging research, please find below a compiled list of recommended IGI Global book chapters and journal articles in the areas of libraries, emergency management, and library staff. These related readings will provide additional information and guidance to further enrich your knowledge and assist you with your own research.

Adams, T. (2014). An Examination of Challenges Faced by First Responders in the Midst of Disaster. In C. Brown, K. Peters, & K. Nyarko (Eds.), *Cases on Research and Knowledge Discovery* (pp. 198–216). Hershey, PA: Homeland Security Centers of Excellence. doi:10.4018/978-1-4666-5946-9.ch009

Adler, M. (2015). Floods Monitoring. In C. Maftei (Ed.), *Extreme Weather and Impacts of Climate Change on Water Resources in the Dobrogea Region* (pp. 312–344). Hershey, PA: IGI Global. doi:10.4018/978-1-4666-8438-6.ch011

Agrawal, P. R. (2016). Google Search: Digging into the Culture of Information Retrieval. In E. de Smet & S. Dhamdhere (Eds.), *E-Discovery Tools and Applications in Modern Libraries* (pp. 210–239). Hershey, PA: IGI Global. doi:10.4018/978-1-5225-0474-0.ch012

Akaichi, J., & Mhadhbi, L. (2016). A Clinical Decision Support System: Ontology-Driven Approach for Effective Emergency Management. In J. Moon & M. Galea (Eds.), *Improving Health Management through Clinical Decision Support Systems* (pp. 270–294). Hershey, PA: IGI Global. doi:10.4018/978-1-4666-9432-3.ch013

Alenzuela, R. (2017). Research, Leadership, and Resource-Sharing Initiatives: The Role of Local Library Consortia in Access to Medical Information. In S. Ram (Ed.), *Library and Information Services for Bioinformatics Education and Research* (pp. 199–211). Hershey, PA: IGI Global. doi:10.4018/978-1-5225-1871-6.ch012

Allison, D. (2017). When Sales Talk Meets Reality: Implementing a Self-Checkout Kiosk. In E. Iglesias (Ed.), *Library Technology Funding, Planning, and Deployment* (pp. 36–54). Hershey, PA: IGI Global. doi:10.4018/978-1-5225-1735-1.ch003

Almagrabi, A., Loke, S. W., & Torabi, T. (2014). Spatial Relations in Contextual Information for Mobile Emergency Messaging. In B. Guo, D. Riboni, & P. Hu (Eds.), *Creating Personal, Social, and Urban Awareness through Pervasive Computing* (pp. 274–298). Hershey, PA: IGI Global. doi:10.4018/978-1-4666-4695-7.ch012

Anbu, J. P. (2017). Information Services to Biomedical Science through Mobile Technology Applications. In S. Ram (Ed.), *Library and Information Services for Bioinformatics Education and Research* (pp. 155–168). Hershey, PA: IGI Global. doi:10.4018/978-1-5225-1871-6.ch008

Anderson, L. R. (2017). Assessing the Interactions between Native American Tribes and the U.S. Government in Homeland Security and Emergency Management Policy. In M. Dawson, D. Kisku, P. Gupta, J. Sing, & W. Li (Eds.), *Developing Next-Generation Countermeasures for Homeland Security Threat Prevention* (pp. 51–71). Hershey, PA: IGI Global. doi:10.4018/978-1-5225-0703-1.ch003

Andersson, D. (2014). An Externalizable Model of Tactical Mission Control for Knowledge Transfer. *International Journal of Information Systems for Crisis Response and Management, 6*(3), 16–37. doi:10.4018/IJISCRAM.2014070102

Annunziato, A., Doherty, B., & Khanh, H. (2014). A Modular Collaborative Web-Based Framework for Humanitarian Crisis Management. In Z. Mahmood (Ed.), *IT in the Public Sphere: Applications in Administration* (pp. 18–46). Hershey, PA: Government, Politics, and Planning. doi:10.4018/978-1-4666-4719-0.ch002

Anwar, J. (2015). Reproductive and Mental Health during Natural Disaster: Implications and Issues for Women in Developing Nations – A Case Example. In M. Sheikh, A. Mahamoud, & M. Househ (Eds.), *Transforming Public Health in Developing Nations* (pp. 265–291). Hershey, PA: IGI Global. doi:10.4018/978-1-4666-8702-8.ch012

Arbanas, S. M., & Arbanas, Ž. (2015). Landslides: A Guide to Researching Landslide Phenomena and Processes. In N. Gaurina-Medjimurec (Ed.), *Handbook of Research on Advancements in Environmental Engineering* (pp. 474–510). Hershey, PA: IGI Global. doi:10.4018/978-1-4666-7336-6.ch017

Arias-Hernandez, R., & Fisher, B. (2014). Designing Visual Analytic Tools for Emergency Operation Centers: A Qualitative Approach. *International Journal of Information Systems for Crisis Response and Management*, *6*(3), 1–15. doi:10.4018/IJISCRAM.2014070101

Arlikatti, S., Kendra, J., & Jennings, E. (2015). Corporate Social Responsibility in Enhancing Disaster Education. In M. Hamner, S. Stovall, D. Taha, & S. Brahimi (Eds.), *Emergency Management and Disaster Response Utilizing Public-Private Partnerships* (pp. 183–194). Hershey, PA: IGI Global. doi:10.4018/978-1-4666-8159-0.ch010

Arlikatti, S., Kendra, J., & Jennings, E. (2015). Corporate Social Responsibility in Enhancing Disaster Education. In M. Hamner, S. Stovall, D. Taha, & S. Brahimi (Eds.), *Emergency Management and Disaster Response Utilizing Public-Private Partnerships* (pp. 183–194). Hershey, PA: IGI Global. doi:10.4018/978-1-4666-8159-0.ch010

Attia, S., Boubetra, A., & Saoud, M. S. (2016). Development of an Emergency Response Management using Mobile Devices for Hospital Infrastructures Affected by Power Grid Failures. *International Journal of Healthcare Information Systems and Informatics*, *11*(1), 36–57. doi:10.4018/IJHISI.2016010103

Bahinipati, C. S., Patnaik, U., & Viswanathan, P. K. (2016). What Causes Economic Losses from Natural Disasters in India? In S. Dinda (Ed.), *Handbook of Research on Climate Change Impact on Health and Environmental Sustainability* (pp. 157–175). Hershey, PA: IGI Global. doi:10.4018/978-1-4666-8814-8.ch008

Bahir, E., & Peled, A. (2015). Keyword Selection Methodology for Identification of Major Events using Social Networks. *International Journal of Information Systems for Crisis Response and Management*, *7*(1), 42–60. doi:10.4018/IJISCRAM.2015010103

Baker, W. (2016). Responding to High-Volume Water Disasters in the Research Library Context. In E. Decker & J. Townes (Eds.), *Handbook of Research on Disaster Management and Contingency Planning in Modern Libraries* (pp. 282–310). Hershey, PA: IGI Global. doi:10.4018/978-1-4666-8624-3.ch013

Balaji, S. (2017). An Overview of Biological Data Mining. In S. Ram (Ed.), *Library and Information Services for Bioinformatics Education and Research* (pp. 130–154). Hershey, PA: IGI Global. doi:10.4018/978-1-5225-1871-6.ch007

Balogun, O., & Tetteh, E. G. (2015). Data Fusion Aiding Tool (DAFAT) Design for Emergency Command and Control Using Lean Principles. In E. Tetteh & B. Uzochukwu (Eds.), *Lean Six Sigma Approaches in Manufacturing* (pp. 202–230). Hershey, PA: Services, and Production. doi:10.4018/978-1-4666-7320-5.ch008

Baroni, P., Fogli, D., Giacomin, M., Guida, G., Provenza, L. P., Rossi, M., & Žnidaršič, M. et al. (2014). A Participatory Approach to Designing Decision Support Systems in Emergency Management. *International Journal of Decision Support System Technology*, *6*(1), 60–80. doi:10.4018/ijdsst.2014010104

Barthe-Delanoë, A., Carbonnel, S., Bénaben, F., & Pingaud, H. (2014). An Event-Driven Platform for Agility Management of Crisis Response. *International Journal of Information Systems for Crisis Response and Management*, *6*(2), 54–70. doi:10.4018/ijiscram.2014040104

Basu, J. P. (2016). Coastal Poverty, Resource-Dependent Livelihood, Climate Change, and Adaptation: An Empirical Study in Indian Coastal Sunderbans. In S. Dinda (Ed.), *Handbook of Research on Climate Change Impact on Health and Environmental Sustainability* (pp. 441–454). Hershey, PA: IGI Global. doi:10.4018/978-1-4666-8814-8.ch022

Behnam, B. (2017). Simulating Post-Earthquake Fire Loading in Conventional RC Structures. In P. Samui, S. Chakraborty, & D. Kim (Eds.), *Modeling and Simulation Techniques in Structural Engineering* (pp. 425–444). Hershey, PA: IGI Global. doi:10.4018/978-1-5225-0588-4.ch015

Behnassi, M., Kahime, K., Boussaa, S., Boumezzough, A., & Messouli, M. (2017). Infectious Diseases and Climate Vulnerability in Morocco: Governance and Adaptation Options. In M. Bouzid (Ed.), *Examining the Role of Environmental Change on Emerging Infectious Diseases and Pandemics* (pp. 138–162). Hershey, PA: IGI Global. doi:10.4018/978-1-5225-0553-2.ch006

Bender, S. R. (2016). Floodplain Infrastructure and the Toxic Tide. In A. McKeown & G. Bugyi (Eds.), *Impact of Water Pollution on Human Health and Environmental Sustainability* (pp. 150–173). Hershey, PA: IGI Global. doi:10.4018/978-1-4666-9559-7.ch007

Bengtson, J. (2017). Funding a Gamification Machine. In E. Iglesias (Ed.), *Library Technology Funding, Planning, and Deployment* (pp. 99–112). Hershey, PA: IGI Global. doi:10.4018/978-1-5225-1735-1.ch006

Berndt, H., Mentler, T., & Herczeg, M. (2015). Optical Head-Mounted Displays in Mass Casualty Incidents: Keeping an Eye on Patients and Hazardous Materials. *International Journal of Information Systems for Crisis Response and Management*, *7*(3), 1–15. doi:10.4018/IJISCRAM.2015070101

Bhebhe, S., & Ngwenya, S. (2016). Adoption and Use of Discovery Tools by Selected Academic Libraries in Zimbabwe. In E. de Smet & S. Dhamdhere (Eds.), *E-Discovery Tools and Applications in Modern Libraries* (pp. 168–180). Hershey, PA: IGI Global. doi:10.4018/978-1-5225-0474-0.ch009

Bimonte, S., Boucelma, O., Machabert, O., & Sellami, S. (2014). A Generic Spatial OLAP Model for Evaluating Natural Hazards in a Volunteered Geographic Information Context. *International Journal of Agricultural and Environmental Information Systems*, *5*(4), 40–55. doi:10.4018/ijaeis.2014100102

Boersma, K., Diks, D., Ferguson, J., & Wolbers, J. (2016). From Reactive to Proactive Use of Social Media in Emergency Response: A Critical Discussion of the Twitcident Project. In G. Silvius (Ed.), *Strategic Integration of Social Media into Project Management Practice* (pp. 236–252). Hershey, PA: IGI Global. doi:10.4018/978-1-4666-9867-3.ch014

Bouzid, M. (2017). Waterborne Diseases and Climate Change: Impact and Implications. In M. Bouzid (Ed.), *Examining the Role of Environmental Change on Emerging Infectious Diseases and Pandemics* (pp. 89–108). Hershey, PA: IGI Global. doi:10.4018/978-1-5225-0553-2.ch004

Bowers, M., & Cherne, G. (2015). A Lessons Framework for Civil-Military-Police Conflict and Disaster Management: An Australian Perspective. In S. McIntyre, K. Dalkir, P. Paul, & I. Kitimbo (Eds.), *Utilizing Evidence-Based Lessons Learned for Enhanced Organizational Innovation and Change* (pp. 152–171). Hershey, PA: IGI Global. doi:10.4018/978-1-4666-6453-1.ch008

Bowles, D. C. (2017). Climate Change-Associated Conflict and Infectious Disease. In M. Bouzid (Ed.), *Examining the Role of Environmental Change on Emerging Infectious Diseases and Pandemics* (pp. 68–88). Hershey, PA: IGI Global. doi:10.4018/978-1-5225-0553-2.ch003

Brengarth, L. B., Mujkic, E., & Millar, M. A. (2015). Social Media in Crisis: How Social Media Created a NPO and Relief during a Wildfire Crisis. In H. Asencio & R. Sun (Eds.), *Cases on Strategic Social Media Utilization in the Nonprofit Sector* (pp. 1–23). Hershey, PA: IGI Global. doi:10.4018/978-1-4666-8188-0.ch001

Cai, T. (2014). Geospatial Technology-Based E-Government Design for Environmental Protection and Emergency Response. In K. Bwalya (Ed.), *Technology Development and Platform Enhancements for Successful Global E-Government Design* (pp. 157–184). Hershey, PA: IGI Global. doi:10.4018/978-1-4666-4900-2.ch009

Carmicheal, D. W. (2016). The Incident Command System: Applying Emergency Response Best Practice to Your Disaster. In E. Decker & J. Townes (Eds.), *Handbook of Research on Disaster Management and Contingency Planning in Modern Libraries* (pp. 25–48). Hershey, PA: IGI Global. doi:10.4018/978-1-4666-8624-3.ch002

Carroll, V. (2016). Conservation Since 2000. In E. Decker & J. Townes (Eds.), *Handbook of Research on Disaster Management and Contingency Planning in Modern Libraries* (pp. 467–493). Hershey, PA: IGI Global. doi:10.4018/978-1-4666-8624-3.ch020

Carvalho, T., Santos, L. B., Luz, E. F., Ishibashi, R., Jorge, A. A., & Londe, L. R. (2016). An Open Source Approach for Watersheds Delimitation to Support Flood Monitoring. *International Journal of Distributed Systems and Technologies, 7*(4), 77–88. doi:10.4018/IJDST.2016100105

Chandler, D. R. (2016). Prepared for Anything and Everything: Libraries, Archives, and Unexpected Small Scale Disasters. In E. Decker & J. Townes (Eds.), *Handbook of Research on Disaster Management and Contingency Planning in Modern Libraries* (pp. 240–256). Hershey, PA: IGI Global. doi:10.4018/978-1-4666-8624-3.ch011

Chandler, T., & Beedasy, J. (2015). mLearning to Enhance Disaster Preparedness Education in K-12 Schools. In H. An, S. Alon, & D. Fuentes (Eds.), Tablets in K-12 Education: Integrated Experiences and Implications (pp. 75-89). Hershey, PA: IGI Global. doi:10.4018/978-1-4666-6300-8.ch006

Chandler, T., Crocco, M., & Marri, A. R. (2014). Systems Thinking about Severe Storms in Social Studies Education. In K. Thomas & H. Muga (Eds.), *Handbook of Research on Pedagogical Innovations for Sustainable Development* (pp. 348–367). Hershey, PA: IGI Global. doi:10.4018/978-1-4666-5856-1.ch016

Chapman, J., Davis, L. B., Samanlioglu, F., & Qu, X. (2014). Evaluating the Effectiveness of Pre-Positioning Policies in Response to Natural Disasters. *International Journal of Operations Research and Information Systems*, *5*(2), 86–100. doi:10.4018/ijoris.2014040105

Chatterjee, T., & Dinda, S. (2016). Climate Change, Human Health and Some Economic Issues. In S. Dinda (Ed.), *Handbook of Research on Climate Change Impact on Health and Environmental Sustainability* (pp. 26–41). Hershey, PA: IGI Global. doi:10.4018/978-1-4666-8814-8.ch002

Chaudron, G. (2016). After the Flood: Lessons Learned from Small-Scale Disasters. In E. Decker & J. Townes (Eds.), *Handbook of Research on Disaster Management and Contingency Planning in Modern Libraries* (pp. 389–411). Hershey, PA: IGI Global. doi:10.4018/978-1-4666-8624-3.ch017

Chen, J., Lan, X., Huang, Q., Dong, J., & Chen, C. (2017). Scholarly Learning Commons. In L. Ruan, Q. Zhu, & Y. Ye (Eds.), *Academic Library Development and Administration in China* (pp. 90–109). Hershey, PA: IGI Global. doi:10.4018/978-1-5225-0550-1.ch006

Chen, W., Yao, F., & Jiang, A. (2017). Technology Innovations in Academic Libraries in China. In L. Ruan, Q. Zhu, & Y. Ye (Eds.), *Academic Library Development and Administration in China* (pp. 196–216). Hershey, PA: IGI Global. doi:10.4018/978-1-5225-0550-1.ch012

Chin, S. (2016). Surviving Sandy: Recovering Collections after a Natural Disaster. In E. Decker & J. Townes (Eds.), *Handbook of Research on Disaster Management and Contingency Planning in Modern Libraries* (pp. 366–388). Hershey, PA: IGI Global. doi:10.4018/978-1-4666-8624-3.ch016

Chtourou, W., & Bouzguenda, L. (2015). Interaction Protocols Adaptation Based Coordination for Crisis Management Processes. *International Journal of Information Systems for Crisis Response and Management*, *7*(4), 67–87. doi:10.4018/IJISCRAM.2015100104

Chun, S. A., & Artigas, F. (2015). Participatory Environmental Planning Platform. In C. Silva (Ed.), *Emerging Issues* (pp. 46–68). Hershey, PA: Challenges, and Opportunities in Urban E-Planning. doi:10.4018/978-1-4666-8150-7.ch003

Çinal, H., Taşkan, Ş., & Baybaş, F. (2015). Dynamic Disaster Coordination System with Web based Html5 API. *International Journal of 3-D Information Modeling (IJ3DIM), 4*(2), 1-15. doi:10.4018/IJ3DIM.2015040101

Confalonieri, U., Menezes, J. A., & Margonari, C. (2017). Environmental Change and the Emergence of Infectious Diseases: A Regional Perspective from South America. In M. Bouzid (Ed.), *Examining the Role of Environmental Change on Emerging Infectious Diseases and Pandemics* (pp. 109–137). Hershey, PA: IGI Global. doi:10.4018/978-1-5225-0553-2.ch005

Cowick, C., & Cowick, J. (2016). Planning for a Disaster: Effective Emergency Management in the 21st Century. In E. Decker & J. Townes (Eds.), *Handbook of Research on Disaster Management and Contingency Planning in Modern Libraries* (pp. 49–69). Hershey, PA: IGI Global. doi:10.4018/978-1-4666-8624-3.ch003

Cui, Y. (2017). Research Data Management: Models, Challenges, and Actions. In L. Ruan, Q. Zhu, & Y. Ye (Eds.), *Academic Library Development and Administration in China* (pp. 184–195). Hershey, PA: IGI Global. doi:10.4018/978-1-5225-0550-1.ch011

Curran, S. J. (2015). Application of Cellular Communications Models and Designs for Use in Disaster-Aftermath Related Scenarios. *International Journal of Interdisciplinary Telecommunications and Networking*, *7*(3), 46–56. doi:10.4018/IJITN.2015070104

de Keyser, P. (2016). Discovery Services, the Next Tool Libraries Must Have, or… Should Discard? In E. de Smet & S. Dhamdhere (Eds.), *E-Discovery Tools and Applications in Modern Libraries* (pp. 56–65). Hershey, PA: IGI Global. doi:10.4018/978-1-5225-0474-0.ch003

de Smet, E. (2016). E-Discovery with the ABCD Information Management System. In E. de Smet & S. Dhamdhere (Eds.), *E-Discovery Tools and Applications in Modern Libraries* (pp. 332–357). Hershey, PA: IGI Global. doi:10.4018/978-1-5225-0474-0.ch017

Dhamdhere, S. N., & Lihitkar, R. (2016). Commercial and Open Access Integrated Information Search Tools in Indian Libraries. In E. de Smet & S. Dhamdhere (Eds.), *E-Discovery Tools and Applications in Modern Libraries* (pp. 41–55). Hershey, PA: IGI Global. doi:10.4018/978-1-5225-0474-0.ch002

Dixon, J., & Abashian, N. (2016). Beyond the Collection: Emergency Planning for Public and Staff Safety. In E. Decker & J. Townes (Eds.), *Handbook of Research on Disaster Management and Contingency Planning in Modern Libraries* (pp. 120–140). Hershey, PA: IGI Global. doi:10.4018/978-1-4666-8624-3.ch006

Egan, J., & Anderson, T. (2015). Considerations for a Model of Public-Private Sector Collaboration in the Provision of Disaster Relief: Incentives and Limits. In M. Hamner, S. Stovall, D. Taha, & S. Brahimi (Eds.), *Emergency Management and Disaster Response Utilizing Public-Private Partnerships* (pp. 1–15). Hershey, PA: IGI Global. doi:10.4018/978-1-4666-8159-0.ch001

Fagbohun, M. O., Nwokocha, N. M., Itsekor, V., & Adebayo, O. (2016). Responsive Library Website Design and Adoption of Federated Search Tools for Library Services in Developing Countries. In E. de Smet & S. Dhamdhere (Eds.), *E-Discovery Tools and Applications in Modern Libraries* (pp. 76–108). Hershey, PA: IGI Global. doi:10.4018/978-1-5225-0474-0.ch005

Fan, Y., Zhang, X., & Li, G. (2017). Research Initiatives and Projects in Academic Libraries. In L. Ruan, Q. Zhu, & Y. Ye (Eds.), *Academic Library Development and Administration in China* (pp. 230–252). Hershey, PA: IGI Global. doi:10.4018/978-1-5225-0550-1.ch014

Fleischer, S. V., & Calzonetti, J. A. (2016). Deferred Maintenance and Space Repurposing: The Impact on Libraries and Archives Disaster and Contingency Planning. In E. Decker & J. Townes (Eds.), *Handbook of Research on Disaster Management and Contingency Planning in Modern Libraries* (pp. 257–281). Hershey, PA: IGI Global. doi:10.4018/978-1-4666-8624-3.ch012

Galloup, A. (2016). One Plan, Four Libraries: A Case Study in Disaster Planning for a Four-Campus Academic Institution. In E. Decker & J. Townes (Eds.), *Handbook of Research on Disaster Management and Contingency Planning in Modern Libraries* (pp. 166–183). Hershey, PA: IGI Global. doi:10.4018/978-1-4666-8624-3.ch008

Gibbons, P. (2016). Disaster Management and Exhibition Loans: Contingency Planning for Items on Display. In E. Decker & J. Townes (Eds.), *Handbook of Research on Disaster Management and Contingency Planning in Modern Libraries* (pp. 141–165). Hershey, PA: IGI Global. doi:10.4018/978-1-4666-8624-3.ch007

Goldman, B. (2016). Two Fires and a Flood: Lasting Impact on a Public Library, Its Staff, and Community. In E. Decker & J. Townes (Eds.), *Handbook of Research on Disaster Management and Contingency Planning in Modern Libraries* (pp. 560–581). Hershey, PA: IGI Global. doi:10.4018/978-1-4666-8624-3.ch024

Goovaerts, M., Nieuwenhuysen, P., & Dhamdhere, S. N. (2016). VLIR-UOS Workshop 'E-Info Discovery and Management for Institutes in the South': Presentations and Conclusions, Antwerp, 8-19 December, 2014. In E. de Smet, & S. Dhamdhere (Eds.), E-Discovery Tools and Applications in Modern Libraries (pp. 1-40). Hershey, PA: IGI Global. doi:10.4018/978-1-5225-0474-0.ch001

Gu, J. (2017). Library Buildings on New Campuses. In L. Ruan, Q. Zhu, & Y. Ye (Eds.), *Academic Library Development and Administration in China* (pp. 110–124). Hershey, PA: IGI Global. doi:10.4018/978-1-5225-0550-1.ch007

Guan, Z., & Wang, J. (2017). The China Academic Social Sciences and Humanities Library (CASHL). In L. Ruan, Q. Zhu, & Y. Ye (Eds.), *Academic Library Development and Administration in China* (pp. 31–54). Hershey, PA: IGI Global. doi:10.4018/978-1-5225-0550-1.ch003

Guo, J., Zhang, H., & Zong, Y. (2017). Leadership Development and Career Planning. In L. Ruan, Q. Zhu, & Y. Ye (Eds.), *Academic Library Development and Administration in China* (pp. 264–279). Hershey, PA: IGI Global. doi:10.4018/978-1-5225-0550-1.ch016

Gupta, P., & Jha, R. S. (2017). Information Retrieval and Access in Cloud. In S. Ram (Ed.), *Library and Information Services for Bioinformatics Education and Research* (pp. 212–228). Hershey, PA: IGI Global. doi:10.4018/978-1-5225-1871-6.ch013

Hahn, D. (2015). Using Systems Analysis and Dynamics to Evaluate Public Private Partnerships with Data and Best Practices from SAFER Santa Rosa. In M. Hamner, S. Stovall, D. Taha, & S. Brahimi (Eds.), *Emergency Management and Disaster Response Utilizing Public-Private Partnerships* (pp. 195–216). Hershey, PA: IGI Global. doi:10.4018/978-1-4666-8159-0.ch011

Hamilton, R., & Brown, D. (2016). Disaster Management and Continuity Planning in Libraries: Changes since the Year 2000. In E. Decker & J. Townes (Eds.), *Handbook of Research on Disaster Management and Contingency Planning in Modern Libraries* (pp. 1–24). Hershey, PA: IGI Global. doi:10.4018/978-1-4666-8624-3.ch001

Hamner, M. (2015). Closing the Gaps in Public Private Partnerships in Emergency Management: A Gap Analysis. In M. Hamner, S. Stovall, D. Taha, & S. Brahimi (Eds.), *Emergency Management and Disaster Response Utilizing Public-Private Partnerships* (pp. 64–97). Hershey, PA: IGI Global. doi:10.4018/978-1-4666-8159-0.ch005

Harris, E. N. (2015). A Public Sector Practitioner's Perspective on Public Private Partnerships. In M. Hamner, S. Stovall, D. Taha, & S. Brahimi (Eds.), *Emergency Management and Disaster Response Utilizing Public-Private Partnerships* (pp. 54–63). Hershey, PA: IGI Global. doi:10.4018/978-1-4666-8159-0.ch004

Harris, E. N. (2015). Coalitions: The Future of Healthcare in Public Private Partnerships. In M. Hamner, S. Stovall, D. Taha, & S. Brahimi (Eds.), *Emergency Management and Disaster Response Utilizing Public-Private Partnerships* (pp. 240–255). Hershey, PA: IGI Global. doi:10.4018/978-1-4666-8159-0.ch013

Harris, E. N. (2015). Trends in Public Private Partnerships. In M. Hamner, S. Stovall, D. Taha, & S. Brahimi (Eds.), *Emergency Management and Disaster Response Utilizing Public-Private Partnerships* (pp. 16–31). Hershey, PA: IGI Global. doi:10.4018/978-1-4666-8159-0.ch002

Hewitt, A. M., Wagner, S. L., Twal, R., & Gourley, D. (2015). Aligning Community Hospitals with Local Public Health Departments: Collaborative Emergency Management. In M. Hamner, S. Stovall, D. Taha, & S. Brahimi (Eds.), *Emergency Management and Disaster Response Utilizing Public-Private Partnerships* (pp. 218–239). Hershey, PA: IGI Global. doi:10.4018/978-1-4666-8159-0.ch012

Huang, C., & Xue, H. F. (2017). The China Academic Digital Associative Library (CADAL). In L. Ruan, Q. Zhu, & Y. Ye (Eds.), *Academic Library Development and Administration in China* (pp. 20–30). Hershey, PA: IGI Global. doi:10.4018/978-1-5225-0550-1.ch002

Huizinga, T., Ayanso, A., Smoor, M., & Wronski, T. (2017). Exploring Insurance and Natural Disaster Tweets Using Text Analytics. *International Journal of Business Analytics, 4*(1), 1–17. doi:10.4018/IJBAN.2017010101

Idiegbeyan-Ose, J., Ifijeh, G., Iwu-James, J., & Ilogho, J. (2016). Management of Institutional Repositories (IR) in Developing Countries. In E. de Smet & S. Dhamdhere (Eds.), *E-Discovery Tools and Applications in Modern Libraries* (pp. 306–331). Hershey, PA: IGI Global. doi:10.4018/978-1-5225-0474-0.ch016

Ifijeh, G., Idiegbeyan-Ose, J., Isiakpona, C. D., & Ilogho, J. (2016). Disaster and Digital Libraries in Developing Countries: Issues and Challenges. In E. Decker & J. Townes (Eds.), *Handbook of Research on Disaster Management and Contingency Planning in Modern Libraries* (pp. 534–559). Hershey, PA: IGI Global. doi:10.4018/978-1-4666-8624-3.ch023

Iglesias, E. (2017). Funding a Makerspace: Making It Up as You Go Along. In E. Iglesias (Ed.), *Library Technology Funding, Planning, and Deployment* (pp. 124–135). Hershey, PA: IGI Global. doi:10.4018/978-1-5225-1735-1.ch008

Iglesias, E. (2017). Insourcing and Outsourcing of Library Technology. In E. Iglesias (Ed.), *Library Technology Funding, Planning, and Deployment* (pp. 113–123). Hershey, PA: IGI Global. doi:10.4018/978-1-5225-1735-1.ch007

Iroaganachi, M. A. (2016). Trends and Issues in Digital Libraries. In E. de Smet & S. Dhamdhere (Eds.), *E-Discovery Tools and Applications in Modern Libraries* (pp. 249–274). Hershey, PA: IGI Global. doi:10.4018/978-1-5225-0474-0.ch014

Jones, A. (2016). Shortcomings and Successes: A Small-Scale Disaster Case Study. In E. Decker & J. Townes (Eds.), *Handbook of Research on Disaster Management and Contingency Planning in Modern Libraries* (pp. 412–435). Hershey, PA: IGI Global. doi:10.4018/978-1-4666-8624-3.ch018

Jumpala, V. V. (2017). Searching Bioinformatics Information Strategies for Effective Use of Search Engine. In S. Ram (Ed.), *Library and Information Services for Bioinformatics Education and Research* (pp. 169–176). Hershey, PA: IGI Global. doi:10.4018/978-1-5225-1871-6.ch009

Kachgal, J. (2015). Maximize Existing Resources with Your Public-Private Partnerships. In M. Hamner, S. Stovall, D. Taha, & S. Brahimi (Eds.), *Emergency Management and Disaster Response Utilizing Public-Private Partnerships* (pp. 270–281). Hershey, PA: IGI Global. doi:10.4018/978-1-4666-8159-0.ch015

Kasemsap, K. (2016). Mastering Digital Libraries in the Digital Age. In E. de Smet & S. Dhamdhere (Eds.), *E-Discovery Tools and Applications in Modern Libraries* (pp. 275–305). Hershey, PA: IGI Global. doi:10.4018/978-1-5225-0474-0.ch015

Kenausis, V., & Herman, D. (2017). Don't Make Us Use the "Get Along Shirt": Communication and Consensus Building in an RFP Process. In E. Iglesias (Ed.), *Library Technology Funding, Planning, and Deployment* (pp. 1–22). Hershey, PA: IGI Global. doi:10.4018/978-1-5225-1735-1.ch001

Khandelwal, I., Sharma, A., Agrawal, P. K., & Shrivastava, R. (2017). Bioinformatics Database Resources. In S. Ram (Ed.), *Library and Information Services for Bioinformatics Education and Research* (pp. 45–90). Hershey, PA: IGI Global. doi:10.4018/978-1-5225-1871-6.ch004

Kohl, L. E., Lombardi, P., & Moroney, M. (2017). Moving from Local to Global via the Integrated Library System: Cost-Savings, ILS Management, Teams, and End-Users. In E. Iglesias (Ed.), *Library Technology Funding, Planning, and Deployment* (pp. 23–35). Hershey, PA: IGI Global. doi:10.4018/978-1-5225-1735-1.ch002

Lock, M. B., Fansler, C., & Webb, M. (2016). Emergency Planning (R)Evolution: Making a Comprehensive Emergency Plan for the Present and the Future. In E. Decker & J. Townes (Eds.), *Handbook of Research on Disaster Management and Contingency Planning in Modern Libraries* (pp. 70–95). Hershey, PA: IGI Global. doi:10.4018/978-1-4666-8624-3.ch004

Long, X., & Yao, B. (2017). The Construction and Development of the Academic Digital Library of Chinese Ancient Collections. In L. Ruan, Q. Zhu, & Y. Ye (Eds.), *Academic Library Development and Administration in China* (pp. 126–135). Hershey, PA: IGI Global. doi:10.4018/978-1-5225-0550-1.ch008

Mabe, M. R. (2016). The Library as Lifeboat. In E. Decker & J. Townes (Eds.), *Handbook of Research on Disaster Management and Contingency Planning in Modern Libraries* (pp. 494–515). Hershey, PA: IGI Global. doi:10.4018/978-1-4666-8624-3.ch021

Magda, B. (2015). Communications and Information Sharing in Public-Private Partnerships: Networking for Emergency Management. In M. Hamner, S. Stovall, D. Taha, & S. Brahimi (Eds.), *Emergency Management and Disaster Response Utilizing Public-Private Partnerships* (pp. 162–182). Hershey, PA: IGI Global. doi:10.4018/978-1-4666-8159-0.ch009

Mane, M. B., & Panage, B. (2016). University Library Portal: A Tool for Web-Enabled Information Services. In E. de Smet & S. Dhamdhere (Eds.), *E-Discovery Tools and Applications in Modern Libraries* (pp. 66–75). Hershey, PA: IGI Global. doi:10.4018/978-1-5225-0474-0.ch004

Manlunching. (2017). Information Needs of Bioinformatics Researchers. In S. Ram (Ed.), *Library and Information Services for Bioinformatics Education and Research* (pp. 34-44). Hershey, PA: IGI Global. doi:10.4018/978-1-5225-1871-6.ch003

Marks, A. B., & Owen, E. (2016). It Is Everywhere: Handling a Mold Outbreak in a Library's High-Density Storage Collection. In E. Decker & J. Townes (Eds.), *Handbook of Research on Disaster Management and Contingency Planning in Modern Libraries* (pp. 311–339). Hershey, PA: IGI Global. doi:10.4018/978-1-4666-8624-3.ch014

Maynor, A. (2016). Response to the Unthinkable: Collecting and Archiving Condolence and Temporary Memorial Materials following Public Tragedies. In E. Decker & J. Townes (Eds.), *Handbook of Research on Disaster Management and Contingency Planning in Modern Libraries* (pp. 582–624). Hershey, PA: IGI Global. doi:10.4018/978-1-4666-8624-3.ch025

Moon, T., & Ko, S. (2015). Compassion Organizing for Public-Private Collaboration in Disaster Management. In M. Hamner, S. Stovall, D. Taha, & S. Brahimi (Eds.), *Emergency Management and Disaster Response Utilizing Public-Private Partnerships* (pp. 99–120). Hershey, PA: IGI Global. doi:10.4018/978-1-4666-8159-0.ch006

Na, L. (2017). Library and Information Science Education and Graduate Programs in Academic Libraries. In L. Ruan, Q. Zhu, & Y. Ye (Eds.), *Academic Library Development and Administration in China* (pp. 218–229). Hershey, PA: IGI Global. doi:10.4018/978-1-5225-0550-1.ch013

Nagarkar, S. P. (2017). Biomedical Librarianship in the Post-Genomic Era. In S. Ram (Ed.), *Library and Information Services for Bioinformatics Education and Research* (pp. 1–17). Hershey, PA: IGI Global. doi:10.4018/978-1-5225-1871-6.ch001

Natarajan, M. (2016). Exploring the E-Discovery Tools on the Use of Library Collections by Users. In E. de Smet & S. Dhamdhere (Eds.), *E-Discovery Tools and Applications in Modern Libraries* (pp. 122–137). Hershey, PA: IGI Global. doi:10.4018/978-1-5225-0474-0.ch007

Nazir, T. (2016). E-Metrics: Tools for Measuring Usage of Electronic Resources. In E. de Smet & S. Dhamdhere (Eds.), *E-Discovery Tools and Applications in Modern Libraries* (pp. 201–209). Hershey, PA: IGI Global. doi:10.4018/978-1-5225-0474-0.ch011

Negron, M. (2015). Use Team Building to Make the Most of Your Public-Private Partnerships. In M. Hamner, S. Stovall, D. Taha, & S. Brahimi (Eds.), *Emergency Management and Disaster Response Utilizing Public-Private Partnerships* (pp. 121–135). Hershey, PA: IGI Global. doi:10.4018/978-1-4666-8159-0.ch007

Negrón, M. A., & Taha, D. (2015). Public-Private Partnerships in Support of Critical Infrastructure and Key Resources. In M. Hamner, S. Stovall, D. Taha, & S. Brahimi (Eds.), *Emergency Management and Disaster Response Utilizing Public-Private Partnerships* (pp. 256–269). Hershey, PA: IGI Global. doi:10.4018/978-1-4666-8159-0.ch014

Nielit, S. G., & S., T. (2016). E-Discovery Components of E-Teaching And M-Learning: An Overview. In E. de Smet, & S. Dhamdhere (Eds.), *E-Discovery Tools and Applications in Modern Libraries* (pp. 240-248). Hershey, PA: IGI Global. doi:10.4018/978-1-5225-0474-0.ch013

Nixon, M. L. (2016). Safety Doesn't Happen by Accident: Disaster Planning at the University of Pittsburgh. In E. Decker & J. Townes (Eds.), *Handbook of Research on Disaster Management and Contingency Planning in Modern Libraries* (pp. 184–206). Hershey, PA: IGI Global. doi:10.4018/978-1-4666-8624-3.ch009

Panigrahi, P. P., & Singh, T. R. (2017). Data Mining, Big Data, Data Analytics: Big Data Analytics in Bioinformatics. In S. Ram (Ed.), *Library and Information Services for Bioinformatics Education and Research* (pp. 91–111). Hershey, PA: IGI Global. doi:10.4018/978-1-5225-1871-6.ch005

Pareek, R., & Kumar, S. (2017). Open Access Journal in Bioinformatics: A Study. In S. Ram (Ed.), *Library and Information Services for Bioinformatics Education and Research* (pp. 229–240). Hershey, PA: IGI Global. doi:10.4018/978-1-5225-1871-6.ch014

Phuritsabam, B., & Devi, A. B. (2017). Information Seeking Behavior of Medical Scientists at Jawaharlal Nehru Institute of Medical Science: A Study. In S. Ram (Ed.), *Library and Information Services for Bioinformatics Education and Research* (pp. 177–187). Hershey, PA: IGI Global. doi:10.4018/978-1-5225-1871-6.ch010

Pionke, J. (2016). Disaster is in the Eye of the Beholder. In E. Decker & J. Townes (Eds.), *Handbook of Research on Disaster Management and Contingency Planning in Modern Libraries* (pp. 516–533). Hershey, PA: IGI Global. doi:10.4018/978-1-4666-8624-3.ch022

Rahman, M. M., & Shoeb, M. Z. (2016). Embracing E-Resource Discovery Technique Using Open Source Software: The IUB Library's Experience. In E. de Smet & S. Dhamdhere (Eds.), *E-Discovery Tools and Applications in Modern Libraries* (pp. 138–167). Hershey, PA: IGI Global. doi:10.4018/978-1-5225-0474-0.ch008

Ram, S. (2017). Library Services for Bioinformatics: Establishing Synergy Data Information and Knowledge. In S. Ram (Ed.), *Library and Information Services for Bioinformatics Education and Research* (pp. 18–33). Hershey, PA: IGI Global. doi:10.4018/978-1-5225-1871-6.ch002

Ram, S. (2017). Web Resources on Zea mays: An Overview. In S. Ram (Ed.), *Library and Information Services for Bioinformatics Education and Research* (pp. 241–252). Hershey, PA: IGI Global. doi:10.4018/978-1-5225-1871-6.ch015

Rao, Y., & Zhang, Y. (2017). The Construction and Development of Academic Library Digital Special Subject Databases. In L. Ruan, Q. Zhu, & Y. Ye (Eds.), *Academic Library Development and Administration in China* (pp. 163–183). Hershey, PA: IGI Global. doi:10.4018/978-1-5225-0550-1.ch010

Reginaldi, J. (2015). Leadership Challenges in Public Private Partnerships in Emergency Management: A Real-World Perspective. In M. Hamner, S. Stovall, D. Taha, & S. Brahimi (Eds.), *Emergency Management and Disaster Response Utilizing Public-Private Partnerships* (pp. 136–161). Hershey, PA: IGI Global. doi:10.4018/978-1-4666-8159-0.ch008

Ruan, L., & Fu, L. (2017). The Trends of Chinese-American Library Relations. In L. Ruan, Q. Zhu, & Y. Ye (Eds.), *Academic Library Development and Administration in China* (pp. 301–333). Hershey, PA: IGI Global. doi:10.4018/978-1-5225-0550-1.ch018

Schuster, D. W. (2017). Selection Process for Free Open Source Software. In E. Iglesias (Ed.), *Library Technology Funding, Planning, and Deployment* (pp. 55–71). Hershey, PA: IGI Global. doi:10.4018/978-1-5225-1735-1.ch004

Sehgal, M., & Singh, T. (2017). Principles and Analysis of Biological Networks: Biological Pathways and Network Motifs. In S. Ram (Ed.), *Library and Information Services for Bioinformatics Education and Research* (pp. 112–129). Hershey, PA: IGI Global. doi:10.4018/978-1-5225-1871-6.ch006

Sha, Y., Hu, W., & Guo, W. (2017). Digital Libraries in the Western Region of China. In L. Ruan, Q. Zhu, & Y. Ye (Eds.), *Academic Library Development and Administration in China* (pp. 136–162). Hershey, PA: IGI Global. doi:10.4018/978-1-5225-0550-1.ch009

Shaw, M. D. (2016). Navigating Campus Disasters from Within the Library: Lessons and Implications from Gulf Coast Institutions. In E. Decker & J. Townes (Eds.), *Handbook of Research on Disaster Management and Contingency Planning in Modern Libraries* (pp. 340–365). Hershey, PA: IGI Global. doi:10.4018/978-1-4666-8624-3.ch015

Silverman, R., Nakashima, T., Hunt, J. M., & Tuia, J. (2016). A Stitch in Time: Disaster Mitigation Strategies for Cultural Heritage Collections. In E. Decker & J. Townes (Eds.), *Handbook of Research on Disaster Management and Contingency Planning in Modern Libraries* (pp. 208–239). Hershey, PA: IGI Global. doi:10.4018/978-1-4666-8624-3.ch010

Sochay, L., & Junus, R. (2017). From Summon to SearchPlus: The RFP Process for a Discovery Tool at the MSU Libraries. In E. Iglesias (Ed.), *Library Technology Funding, Planning, and Deployment* (pp. 72–98). Hershey, PA: IGI Global. doi:10.4018/978-1-5225-1735-1.ch005

Sonawane, M. (2016). Creating an Agile Library. In E. de Smet & S. Dhamdhere (Eds.), *E-Discovery Tools and Applications in Modern Libraries* (pp. 109–121). Hershey, PA: IGI Global. doi:10.4018/978-1-5225-0474-0.ch006

Sorokhaibam, S. D., & Mathabela, N. N. (2017). Information Needs and Assessment of Bioinformatics Students at the University of Swaziland: Librarian View. In S. Ram (Ed.), *Library and Information Services for Bioinformatics Education and Research* (pp. 188–198). Hershey, PA: IGI Global. doi:10.4018/978-1-5225-1871-6.ch011

Taha, D. (2015). Mitigating the Impact of Extreme Events: A Private Sector Perspective on the Value of Public Private Partnerships. In M. Hamner, S. Stovall, D. Taha, & S. Brahimi (Eds.), *Emergency Management and Disaster Response Utilizing Public-Private Partnerships* (pp. 32–53). Hershey, PA: IGI Global. doi:10.4018/978-1-4666-8159-0.ch003

Taha, D. (2015). Mitigating the Impact of Extreme Events: A Private Sector Perspective on the Value of Public Private Partnerships. In M. Hamner, S. Stovall, D. Taha, & S. Brahimi (Eds.), *Emergency Management and Disaster Response Utilizing Public-Private Partnerships* (pp. 32–53). Hershey, PA: IGI Global. doi:10.4018/978-1-4666-8159-0.ch003

Takayama, Y., & Miwa, H. (2015). Delay Tolerant Navigation Method for Fast Evacuation in Poor Communication Environment at the Time of Disaster. *International Journal of Distributed Systems and Technologies*, *6*(3), 29–50. doi:10.4018/IJDST.2015070103

Tang, Q., Xie, Y., & Su, R. (2017). Assessment and Evaluation of Academic Libraries in China. In L. Ruan, Q. Zhu, & Y. Ye (Eds.), *Academic Library Development and Administration in China* (pp. 280–299). Hershey, PA: IGI Global. doi:10.4018/978-1-5225-0550-1.ch017

Tapia, A. H., Giacobe, N. A., Soule, P. J., & LaLone, N. J. (2016). Scaling 911 Texting for Large-Scale Disasters: Developing Practical Technical Innovations for Emergency Management at Public Universities. *International Journal of Public Administration in the Digital Age*, *3*(3), 73–85. doi:10.4018/IJPADA.2016070105

Treurniet, W. (2014). Shaping Comprehensive Emergency Response Networks. In T. Grant, R. Janssen, & H. Monsuur (Eds.), *Network Topology in Command and Control: Organization* (pp. 26–48). Hershey, PA: IGI Global. doi:10.4018/978-1-4666-6058-8.ch002

Trnka, J., & Johansson, B. J. (2014). Resilient Emergency Response: Supporting Flexibility and Improvisation in Collaborative Command and Control. In Crisis Management: Concepts, Methodologies, Tools, and Applications (pp. 813-838). Hershey, PA: IGI Global. doi:10.4018/978-1-4666-4707-7.ch040

Udina, R. (2016). Quality Control and a Method for the Drying of Flooded Archive Collections: The Case of the Municipal Archive of Barcelona (2013). In E. Decker & J. Townes (Eds.), *Handbook of Research on Disaster Management and Contingency Planning in Modern Libraries* (pp. 436–465). Hershey, PA: IGI Global. doi:10.4018/978-1-4666-8624-3.ch019

Vandevoordt, R. (2016). Humanitarian Media Events: On the Symbolic Conditions of Moral Integration. In A. Fox (Ed.), *Global Perspectives on Media Events in Contemporary Society* (pp. 90–105). Hershey, PA: IGI Global. doi:10.4018/978-1-4666-9967-0.ch007

Verplaetse, A., Mascareñas, P., & O'Neill, K. (2016). Zen and the Art of Disaster Planning: Collaboration Challenges in Library Disaster Plan Design and Execution. In E. Decker & J. Townes (Eds.), *Handbook of Research on Disaster Management and Contingency Planning in Modern Libraries* (pp. 96–119). Hershey, PA: IGI Global. doi:10.4018/978-1-4666-8624-3.ch005

Vezyridis, P., Timmons, S., & Wharrad, H. (2016). Lessons Learned from the Implementation of an Emergency Department Information System. In T. Iyamu & A. Tatnall (Eds.), *Maximizing Healthcare Delivery and Management through Technology Integration* (pp. 237–256). Hershey, PA: IGI Global. doi:10.4018/978-1-4666-9446-0.ch014

Vorraber, W., Lichtenegger, G., Brugger, J., Gojmerac, I., Egly, M., Panzenböck, K., & Voessner, S. et al. (2016). Designing Information Systems to Facilitate Civil-Military Cooperation in Disaster Management. *International Journal of Distributed Systems and Technologies*, *7*(4), 22–40. doi:10.4018/IJDST.2016100102

Wridt, P., Seley, J. E., Fisher, S., & DuBois, B. (2014). Participatory Mapping Approaches to Coordinate the Emergency Response of Spontaneous Volunteers After Hurricane Sandy. *International Journal of E-Planning Research*, *3*(3), 1–19. doi:10.4018/ijepr.2014070101

Xiao, L., & Liu, Y. (2017). Development of Innovative User Services. In L. Ruan, Q. Zhu, & Y. Ye (Eds.), *Academic Library Development and Administration in China* (pp. 56–73). Hershey, PA: IGI Global. doi:10.4018/978-1-5225-0550-1.ch004

Xin, X., & Wu, X. (2017). The Practice of Outreach Services in Chinese Special Libraries. In L. Ruan, Q. Zhu, & Y. Ye (Eds.), *Academic Library Development and Administration in China* (pp. 74–89). Hershey, PA: IGI Global. doi:10.4018/978-1-5225-0550-1.ch005

Yao, X., Zhu, Q., & Liu, J. (2017). The China Academic Library and Information System (CALIS). In L. Ruan, Q. Zhu, & Y. Ye (Eds.), *Academic Library Development and Administration in China* (pp. 1–19). Hershey, PA: IGI Global. doi:10.4018/978-1-5225-0550-1.ch001

Yildiz, M., & Demirhan, K. (2016). Analysis and Comparison of the Role of Local Governments with Other Policy Actors in Disaster Relief via Social Media: The Case of Turkey. In U. Sadioglu & K. Dede (Eds.), *Theoretical Foundations and Discussions on the Reformation Process in Local Governments* (pp. 462–483). Hershey, PA: IGI Global. doi:10.4018/978-1-5225-0317-0.ch019

Yin, Q., Yingying, W., Yan, Z., & Xiaojia, M. (2017). Resource Sharing and Mutually Beneficial Cooperation: A Look at the New United Model in Public and College Libraries. In L. Ruan, Q. Zhu, & Y. Ye (Eds.), *Academic Library Development and Administration in China* (pp. 334–352). Hershey, PA: IGI Global. doi:10.4018/978-1-5225-0550-1.ch019

Young-McLear, K., Mazzuchi, T. A., & Sarkani, S. (2015). Large-Scale Disaster Response Management: Social Media and Homeland Security. In J. Sahlin (Ed.), *Social Media and the Transformation of Interaction in Society* (pp. 93–131). Hershey, PA: IGI Global. doi:10.4018/978-1-4666-8556-7.ch006

Yuvaraj, M. (2016). Impact of Discovery Layers on Accessing E-Resources in Academic Libraries: A Case Study of Central University of Bihar. In E. de Smet & S. Dhamdhere (Eds.), *E-Discovery Tools and Applications in Modern Libraries* (pp. 181–200). Hershey, PA: IGI Global. doi:10.4018/978-1-5225-0474-0.ch010

Zhu, S., & Shi, W. (2017). A Bibliometric Analysis of Research and Services in Chinese Academic Libraries. In L. Ruan, Q. Zhu, & Y. Ye (Eds.), *Academic Library Development and Administration in China* (pp. 253–262). Hershey, PA: IGI Global. doi:10.4018/978-1-5225-0550-1.ch015

About the Authors

Michael R. Mabe is the Executive Director of the Chesterfield County Public Library in Chesterfield County, VA. Mike has worked in a variety of executive managerial positions include Assistant Director, Dauphin County Library System, Harrisburg, PA; Dean of Library Services, San Juan College, Farmington, NM and Department Manager, Salt Lake City Public Library, Salt Lake City, UT. He holds a BS from the University of Utah; MPA from Brigham Young University; and an MLIS Brigham Young University. Mike has presented at state and national associations on a variety of topics including, quality management, strategic planning, emergency services, staff training and library innovation. He has also published in the areas emergency services in public libraries and strategic planning. Mike and his wife are the parents of five children and 20 grandchildren. He enjoys woodworking, writing, travelling and being with his grandkids.

Emily Ashley is the Emergency Management Coordinator for Chesterfield County, Virginia. In this role she has served on the Virginia Emergency Management Association Certification Committee and Chair of the Central Virginia Emergency Management Alliance. In 2016 members of Chesterfield County's Mass Care Committee won a National Association of Counties award for the use of libraries and staff post disaster. Before coming to Chesterfield in August of 2011, Emily spent 8 years in various roles of Emergency Management in the Hampton Roads, Virginia. Emily has especially enjoyed emergency management projects related to special needs, catastrophic planning, continuity planning, and mitigation. Emily holds a Bachelor's Degree in Geography from Missouri State University, Master's Degree in Public Administration from Old Dominion University and professional certification as a Virginia Professional Emergency Manager.

Index

H

I

J

L

M

N

O

P

R

S

T

U

V

Purchase Print + Free E-Book or E-Book Only*

Purchase a print book through the IGI Global Online Bookstore and receive the e-book for free or purchase the e-book only! Shipping fees apply.

www.igi-global.com

Recommended Reference Books

ISBN: 978-1-4666-7456-1
© 2015; 2,072 pp.
List Price: $1,880

ISBN: 978-1-4666-5864-6
© 2014; 570 pp.
List Price: $276

ISBN: 978-1-4666-5198-2
© 2014; 398 pp.
List Price: $164

ISBN: 978-1-4666-4679-7
© 2014; 387 pp.
List Price: $140

ISBN: 978-1-4666-6567-5
© 2015; 461 pp.
List Price: $160

ISBN: 978-1-4666-6493-7
© 2015; 643 pp.
List Price: $180

*IGI Global now offers the exclusive opportunity to receive a free e-book with the purchase of the publication in print, or purchase any e-book publication only. You choose the format that best suits your needs. This offer is only valid on purchases made directly through IGI Global's Online Bookstore and not intended for use by book distributors or wholesalers. Shipping fees will be applied for hardcover purchases during checkout if this option is selected.

Should a new edition of any given publication become available, access will not be extended on the new edition and will only be available for the purchased publication. If a new edition becomes available, you will not lose access, but you would no longer receive new content for that publication (i.e. updates). The free e-book is only available to single institutions that purchase printed publications through IGI Global. Sharing the free e-book is prohibited and will result in the termination of e-access.

Publishing Information Science and Technology Research Since 1988

IGI Global — Disseminator of Knowledge

www.igi-global.com | Sign up at www.igi-global.com/newsletters | facebook.com/igiglobal | twitter.com/igiglobal

Stay Current on the Latest Emerging Research Developments

Become an IGI Global Reviewer for Authored Book Projects

The overall success of an authored book project is dependent on quality and timely reviews.

In this competitive age of scholarly publishing, constructive and timely feedback significantly decreases the turnaround time of manuscripts from submission to acceptance, allowing the publication and discovery of progressive research at a much more expeditious rate. Several IGI Global authored book projects are currently seeking highly qualified experts in the field to fill vacancies on their respective editorial review boards:

Applications may be sent to:
development@igi-global.com

Applicants must have a doctorate (or an equivalent degree) as well as publishing and reviewing experience. Reviewers are asked to write reviews in a timely, collegial, and constructive manner. All reviewers will begin their role on an ad-hoc basis for a period of one year, and upon successful completion of this term can be considered for full editorial review board status, with the potential for a subsequent promotion to Associate Editor.

If you have a colleague that may be interested in this opportunity, we encourage you to share this information with them.

www.igi-global.com

InfoSci®-Books

A Database for Progressive Information Science and Technology Research

Maximize Your Library's Book Collection!

Invest in IGI Global's InfoSci®-Books database and gain access to hundreds of reference books at a fraction of their individual list price.

The InfoSci®-Books database offers unlimited simultaneous users the ability to precisely return search results through more than 75,000 full-text chapters from nearly 3,400 reference books in the following academic research areas:

Business & Management Information Science & Technology • Computer Science & Information Technology
Educational Science & Technology • Engineering Science & Technology • Environmental Science & Technology
Government Science & Technology • Library Information Science & Technology • Media & Communication Science & Technology
Medical, Healthcare & Life Science & Technology • Security & Forensic Science & Technology • Social Sciences & Online Behavior

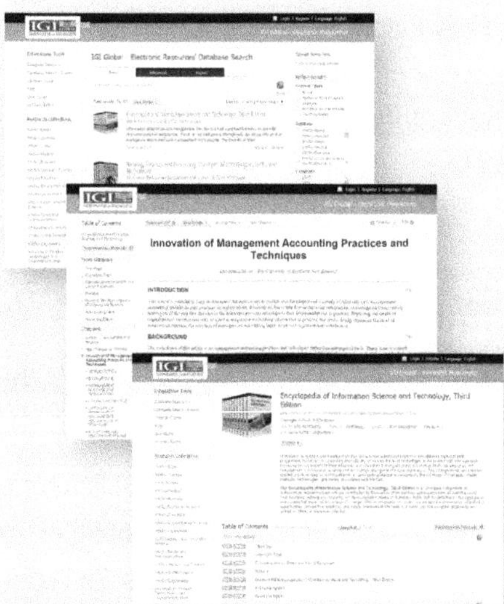

Peer-Reviewed Content:
- Cutting-edge research
- No embargoes
- Scholarly and professional
- Interdisciplinary

Award-Winning Platform:
- Unlimited simultaneous users
- Full-text in XML and PDF
- Advanced search engine
- No DRM

Librarian-Friendly:
- Free MARC records
- Discovery services
- COUNTER4/SUSHI compliant
- Training available

To find out more or request a free trial, visit: www.igi-global.com/eresources

IGI GLOBAL DISSEMINATOR OF KNOWLEDGE
www.igi-global.com

IGI Global
Proudly Partners with

eContent Pro

eContent Pro specializes in the following areas:

Academic Copy Editing
Our expert copy editors will conduct a full copy editing procedure on your manuscript and will also address your preferred reference style to make sure your paper meets the standards of the style of your choice.

Expert Translation
Our expert translators will work to ensure a clear cut and accurate translation of your document, ensuring that your research is flawlessly communicated to your audience.

Professional Proofreading
Our editors will conduct a comprehensive assessment of your content and address all shortcomings of the paper in terms of grammar, language structures, spelling, and formatting.

IGI Global Authors, Save 10% on eContent Pro's Services!

Scan the QR Code to Receive Your 10% Discount

The 10% discount is applied directly to your eContent Pro shopping cart when placing an order through IGI Global's referral link. Use the QR code to access this referral link. eContent Pro has the right to end or modify any promotion at any time.

Email: customerservice@econtentpro.com

econtentpro.com

IRMA INTERNATIONAL

Information Resources Management Association

Become an IRMA Member

Members of the **Information Resources Management Association (IRMA)** understand the importance of community within their field of study. The Information Resources Management Association is an ideal venue through which professionals, students, and academicians can convene and share the latest industry innovations and scholarly research that is changing the field of information science and technology. Become a member today and enjoy the benefits of membership as well as the opportunity to collaborate and network with fellow experts in the field.

IRMA Membership Benefits:

- **One FREE Journal Subscription**
- **30% Off Additional Journal Subscriptions**
- **20% Off Book Purchases**
- Updates on the latest events and research on Information Resources Management through the IRMA-L listserv.
- Updates on new open access and downloadable content added to Research IRM.
- A copy of the Information Technology Management Newsletter twice a year.
- A certificate of membership.

IRMA Membership $195

Scan code or visit **irma-international.org** and begin by selecting your free journal subscription.

Membership is good for one full year.

www.irma-international.org

Encyclopedia of Information Science and Technology, Third Edition (10 Vols.)

Mehdi Khosrow-Pour, D.B.A. (Information Resources Management Association, USA)
ISBN: 978-1-4666-5888-2; **EISBN:** 978-1-4666-5889-9; © 2015; 10,384 pages.

The **Encyclopedia of Information Science and Technology, Third Edition** is a 10-volume compilation of authoritative, previously unpublished research-based articles contributed by thousands of researchers and experts from all over the world. This discipline-defining encyclopedia will serve research needs in numerous fields that are affected by the rapid pace and substantial impact of technological change. With an emphasis on modern issues and the presentation of potential opportunities, prospective solutions, and future directions in the field, it is a relevant and essential addition to any academic library's reference collection.

[1] 30% discount offer cannot be combined with any other discount and is only valid on purchases made directly through IGI Global's Online Bookstore (www.igi-global.com/books), not intended for use by distributors or wholesalers. Offer expires December 31, 2016.

Free Lifetime E-Access with Print Purchase

Take 30% Off Retail Price:

Hardcover with Free E-Access:[2] $2,765
List Price: $3,950

E-Access with Free Hardcover:[2] $2,765
List Price: $3,950

E-Subscription Price:

One (1) Year E-Subscription: $1,288
List Price: $1,840

Two (2) Year E-Subscription: $2,177
List Price: $3,110

Recommend this Title to Your Institution's Library: www.igi-global.com/books

[2] IGI Global now offers the exclusive opportunity to receive free lifetime e-access with the purchase of the publication in print, or purchase any e-access publication and receive a free print copy of the publication. You choose the format that best suits your needs. This offer is only valid on purchases made directly through IGI Global's Online Bookstore and not intended for use by book distributors or wholesalers. Shipping fees will be applied for hardcover purchases during checkout if this option is selected.

The lifetime of a publication refers to its status as the current edition. Should a new edition of any given publication become available, access will not be extended on the new edition and will only be available for the purchased publication. If a new edition becomes available, you will not lose access, but you would no longer receive new content for that publication (i.e. updates). Free Lifetime E-Access is only available to single institutions that purchase printed publications through IGI Global. Sharing the Free Lifetime E-Access is prohibited and will result in the termination of e-access.

www.ingramcontent.com/pod-product-compliance
Ingram Content Group UK Ltd.
Pitfield, Milton Keynes, MK11 3LW, UK
UKHW050944290926
15470UKWH00005B/1633

9 781522 521969